student study

ART NOTEBOOK

BIOLOGY

Third Edition

Peter H. Raven
Director, Missouri Botanical Garden;
Engelmann Professor of Botany,
Washington University, St. Louis, Missouri

George B. Johnson
Professor of Biology,
Washington University, St. Louis, Missouri

WCB Wm. C. Brown Publishers
Dubuque, Iowa • Melbourne, Australia • Oxford, England

Wm. C. Brown Communications, Inc.

President and Chief Executive Officer *G. Franklin Lewis*
Corporate Senior Vice President, President of WCB Manufacturing *Roger Meyer*
Corporate Senior Vice President and Chief Financial Officer *Robert Chesterman*

A Times Mirror Company

ISBN 0–697–24252–8

Printed in the United States of America by Wm. C. Brown Communications, Inc., 2460 Kerper Boulevard, Dubuque, IA 52001

10 9 8 7 6 5 4

TO THE STUDENT

This Student Study Art Notebook is designed to help you in your study of biology. It contains art taken directly from the text and enables you to take notes during lectures, or jot down comments as you are reading through the chapters.

This notebook is perforated and three-hole punched so that you can remove sheets and put them in a binder with other study or lecture notes. Any blank pages at the end of this notebook can be used for additional notes or drawings.

We hope this notebook, used in conjunction with your text, will help make the study of biology easier for you.

DIRECTORY OF NOTEBOOK FIGURES

TO ACCOMPANY RAVEN & JOHNSON, *BIOLOGY*, 3/e

Chapter 2
Atoms Figure 2–2 **1**
Atomic Energy Levels Figure 2–5 **1**
Electron Energy Levels for Helium and Nitrogen Figure 2–7 **2**
Formation of Ionic Bonds by Sodium Chloride Figure 2–9 **2**
Structure of a Hydrogen Bond Figure 2–14 **3**
Why Salt Dissolves in Water Figure 2–18 **3**

Chapter 3
Saturated and Unsaturated Fats Figure 3–12 **4**
Peptide Bond Figure 3–17 **4**
DNA Double Helix Figure 3–23 **5**

Chapter 4
Miller-Urey Experiment Figure 4–3 **5**

Chapter 5
Structure of a Bacterial Cell Figure 5–6 **6**
Animal Cell Figure 5–8 **6**
Plant Cell Figure 5–9 **7**
Rough Endoplasmic Reticulum Figure 5–11 **8**
Golgi Body Figure 5–17 **8**
Nucleus Figure 5–13 **9**
How Proteins Are Secreted Across Membranes Figure 5–18 **10**
Mitochondria in Longitudinal and Cross Section Figure 5–20 **10**
Chloroplast Structure Figure 5–22 **11**
Structure of a Flagellum Figure 5–28 **11**
Centrioles Figure 5–23 **12**

Chapter 6
Cells Have Complex Surfaces Figure 6–5 **13**
Functions of Plasma Membrane Proteins Figure 6–8 **14**
Osmosis Figure 6–11 **15**
How Solutes Create Osmotic Pressure Figure 6–12 **15**
Sodium-Potassium Pump Figure 6–20 **16**

Chapter 7
Energy in Chemical Reactions Figure 7–7 **17**
Exergonic Reactions Figure 7–8 **18**
Catalytic Cycle of an Enzyme Figure 7–10 **19**
Overview of Aerobic Metabolism Figure 7–18 **19**

Chapter 8
How Coupled Reactions Work Figure 8–1 **20**
Chemiosmosis Figure 8–3 **20**
Overview of Cellular Respiration Figure 8–4 **21**
Electron Transport Chain Figure 8–14 **22**
Proton Movement in the Mitochondrion Figure 8–15 **22**

Chapter 9
Path of an Electron in Bacterial Photosynthesis Figure 9–7 **23**
Path of an Electron in a Chloroplast Figure 9–8 **23**
Photosynthetic Electron Transport System Figure 9–9 **24**
Journey into a Leaf Figure 9–14 **24**
Metabolic Machine Figure 9–21 **25**

Chapter 10
Levels of Chromosomal Organization Figure 10–4 **26**

Chapter 11
Sexual Life Cycle Figure 11–4 **27**
Presence of Chiasmata in Reduction Division in Meiosis Figure 11–7 **27**
Meiosis Figure 11–9 **28**
Comparison of Meiosis and Mitosis Figure 11–11 **29**

Chapter 12
Mendel's Cross of Peas Differing in Flower Color Figure 12–13 **30**
Testcross Figure 12–15 **30**
Incomplete Dominance Figure 12–24 **31**

Chapter 13
Amniocentesis Figure 13–14 **31**

Chapter 14
Base Pairing Figure 14–16 **32**
DNA Replication Fork Figure 14–18 **33**

Chapter 15
Transcription Figure 15–4 **34**
Overview of Protein Synthesis Figure 15–5 **35**
Structure of a tRNA Molecule Figure 15–8 **36**
Formation of the Initiation Complex Figure 15–10 **36**
How Polypeptide Synthesis Proceeds Figure 15–11 **37**
Termination of Protein Synthesis Figure 15–13 **37**
How the *lac* Operon Works Figure 15–19 **38**

Chapter 16
Making a Thymine Dimer Figure 16–2 **38**
Slipped Mispairing Figure 16–6 **39**
Portrait of a Cancer Figure 16–8 **39**
How a Mutation Can Cause Cancer Figure 16–12 **40**

Chapter 18
Polymerase Chain Reaction Figure 18–8 **40**

Chapter 19
Hardy-Weinberg Equilibrium Figure 19–3 **41**

Chapter 22
Evolution of Living Hominoids Figure 22–22 **41**

Chapter 23
Competition Among Barnacles Figure 23–12 **42**

Chapter 25
Water Cycle Figure 25–2 **42**
Carbon Cycle Figure 25–3 **43**
Phosphorus Cycle Figure 25–8 **43**
Nitrogen Cycle Figure 25–4 **44**

Chapter 28
Hierarchical System Used in Classifying Organisms Figure 28–4 **45**

Chapter 29
HIV Infection Cycle Figure 29–8 **45**
Lytic and Lysogenic Cycles of a Bacteriophage Figure 29–10 **46**

Chapter 35
Dicot Plant Body Figure 35–1 **47**

Chapter 36
Flow of Plant Materials Figure 36–1 **48**
Pathways of Mineral Transport in Roots Figure 36–4 **48**

Chapter 39
Basic Sponge Structure Figure 39–4 **49**
Basic *Hydra* Structure Figure 39–9 **49**

Chapter 40
Earthworm Figure 40–16 **50**

Chapter 41
Grasshopper Figure 41–7 **50**

Chapter 42
Echinoderm Structure Figure 42–2 **51**

Chapter 43
Idealized Structure of a Vertebrate Neuron Figure 43–13 **51**

Chapter 44
Synaptic Cleft Between Two Neurons Figure 44–13 **52**

Chapter 45
Section Through the Human Brain Figure 45–8 **53**
Monosynaptic Reflexes Figure 45–16 **54**

Chapter 46
Human Ear Structure Figure 46–14 **53**
Human Eye Structure Figure 46–17 **55**
Retina Structure Figure 46–19 **55**

Chapter 47
How Steroid Hormones Work Figure 47–6 **56**
How Peptide Hormones Work Figure 47–7 **57**
Role of the Pituitary Figure 47–11 **57**

Chapter 48
Organization of Compact Bone Figure 48–2 **58**
Mechanisms of Myofilament Contraction Figure 48–13 **59**

Chapter 49
Regulation of Digestive Enzyme Production Figure 49–11 **60**

Chapter 50
Respiratory Journey Figure 50–19 **61**

Chapter 51
Path of Blood Through the Heart Figure 51–20 **62**

Chapter 52
Structure of an Antibody Molecule Figure 52–25 **63**

Chapter 53
Flow of Materials in the Human Kidney Figure 53–9 **64**

Chapter 54
Male Reproductive Organs Figure 54–6 **65**
Female Reproductive Organs Figure 54–11 **65**

Chapter 55
Mammalian Gastrulation Figure 55–12 **66**

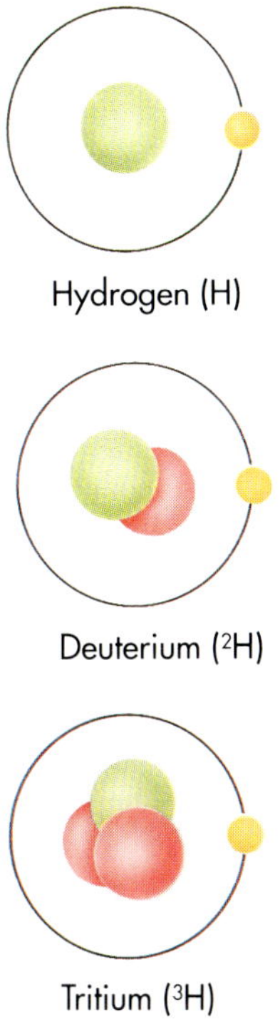

Atoms
Figure 2–2

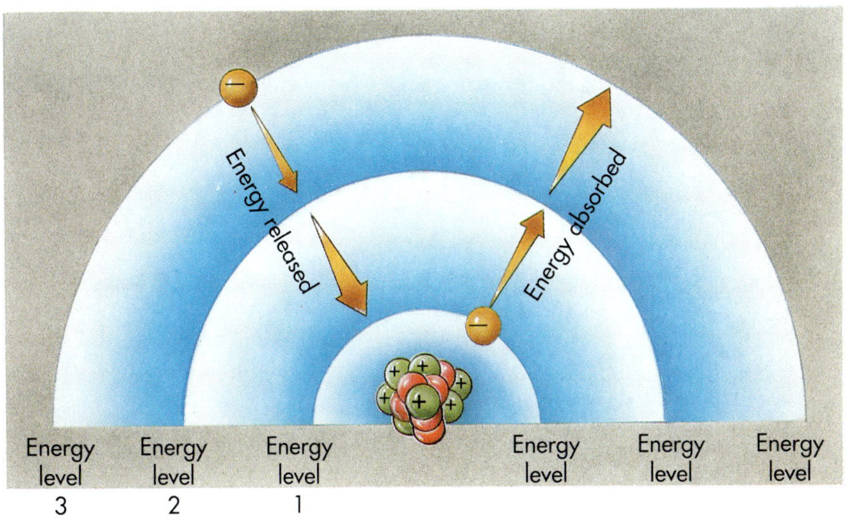

Atomic Energy Levels
Figure 2–5

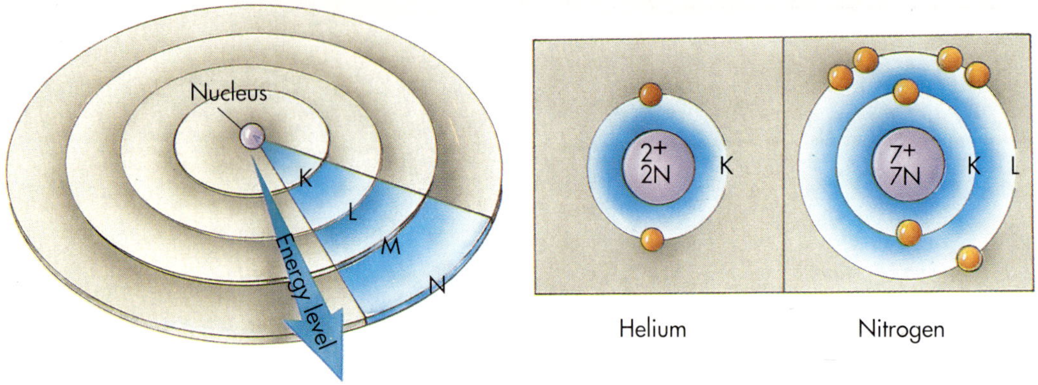

Electron Energy Levels for Helium and Nitrogen
Figure 2–7

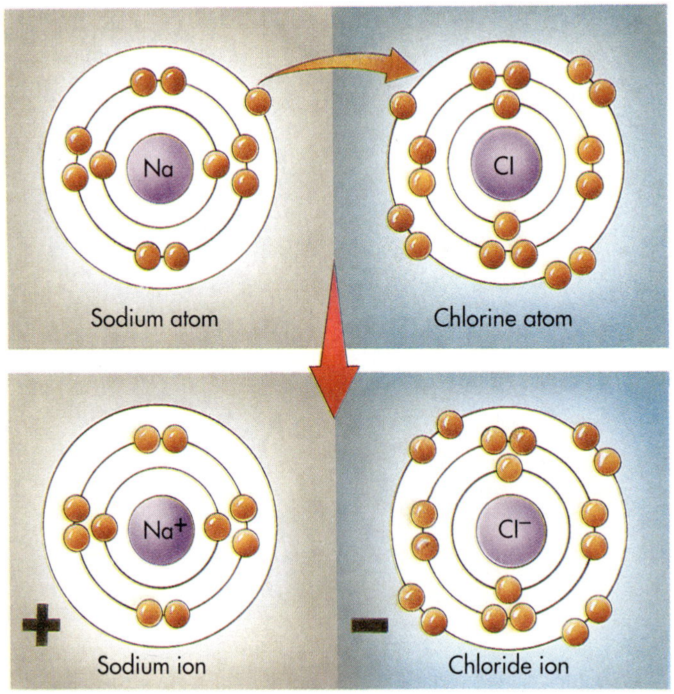

Formation of Ionic Bonds by Sodium Chloride
Figure 2–9

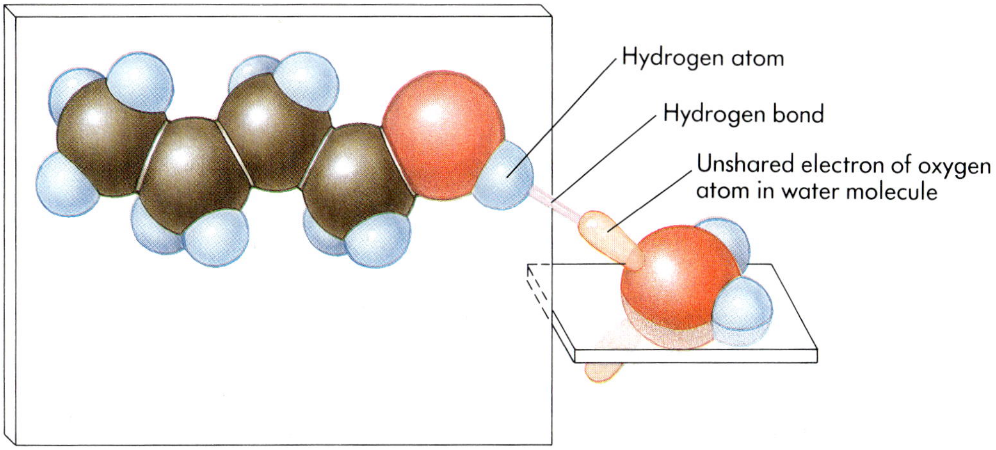

Structure of a Hydrogen Bond
Figure 2–14

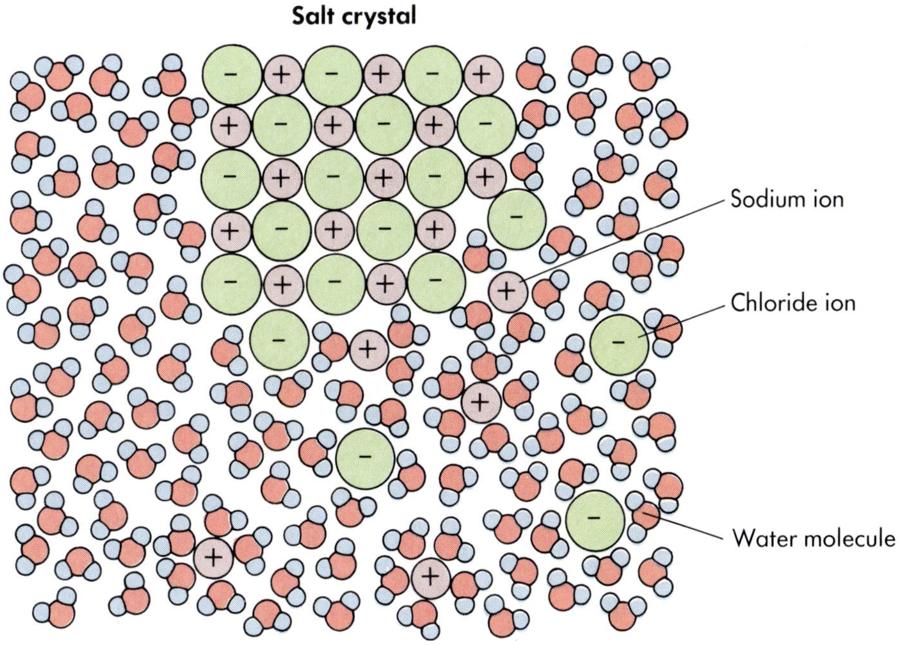

Why Salt Dissolves in Water
Figure 2–18

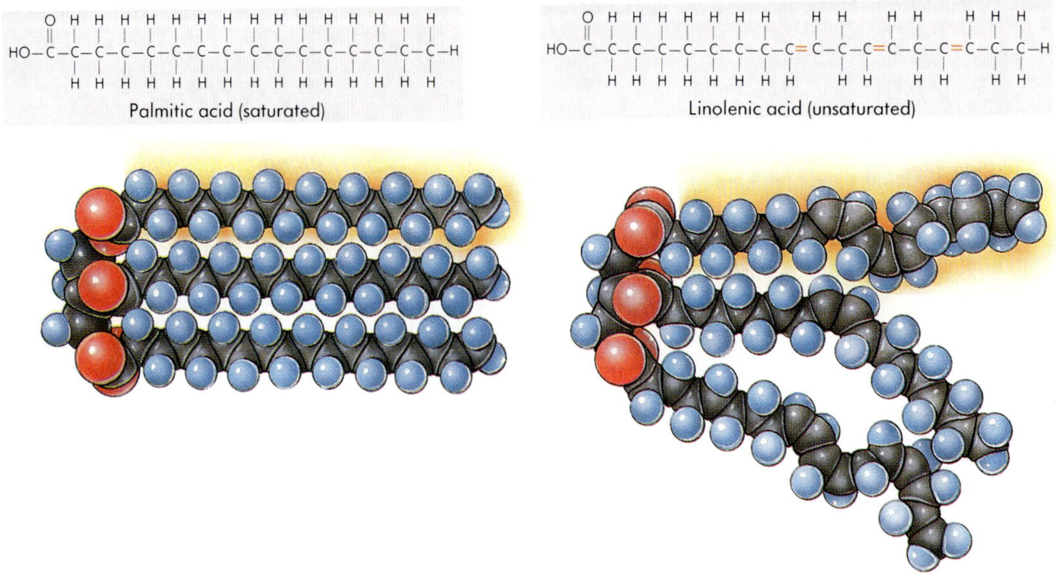

Saturated and Unsaturated Fats
Figure 3–12

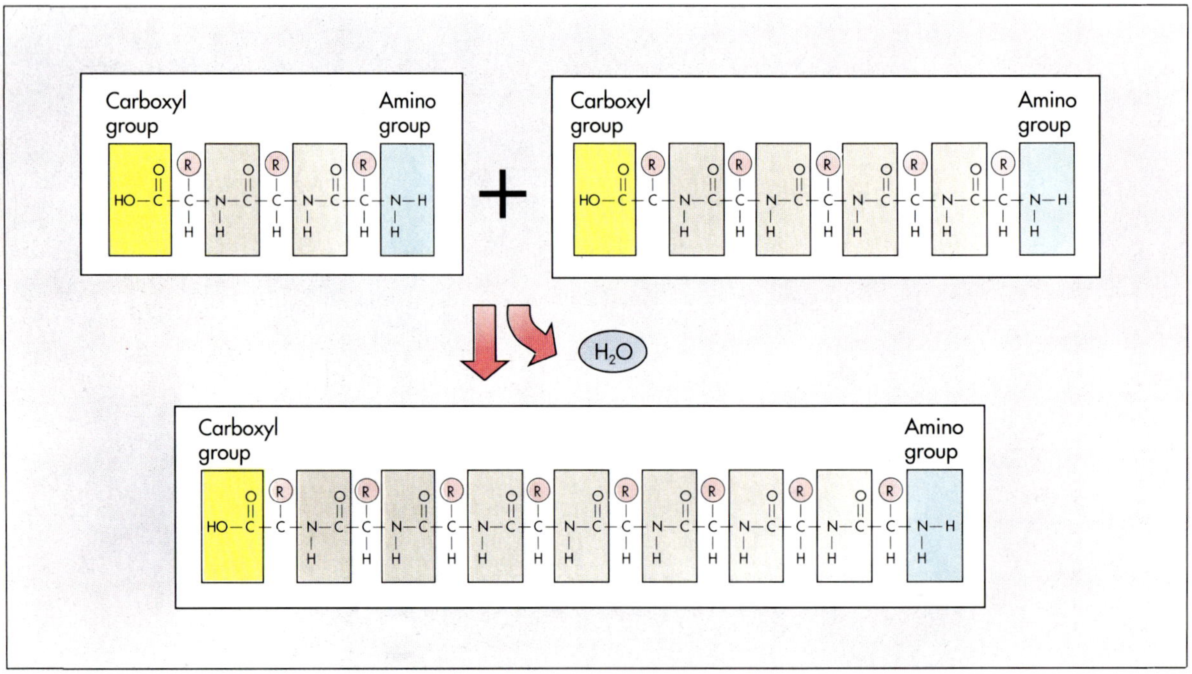

Peptide Bond
Figure 3–17

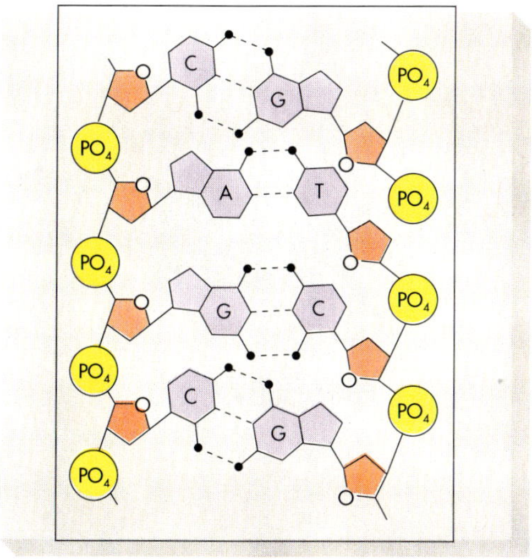

DNA Double Helix
Figure 3–23

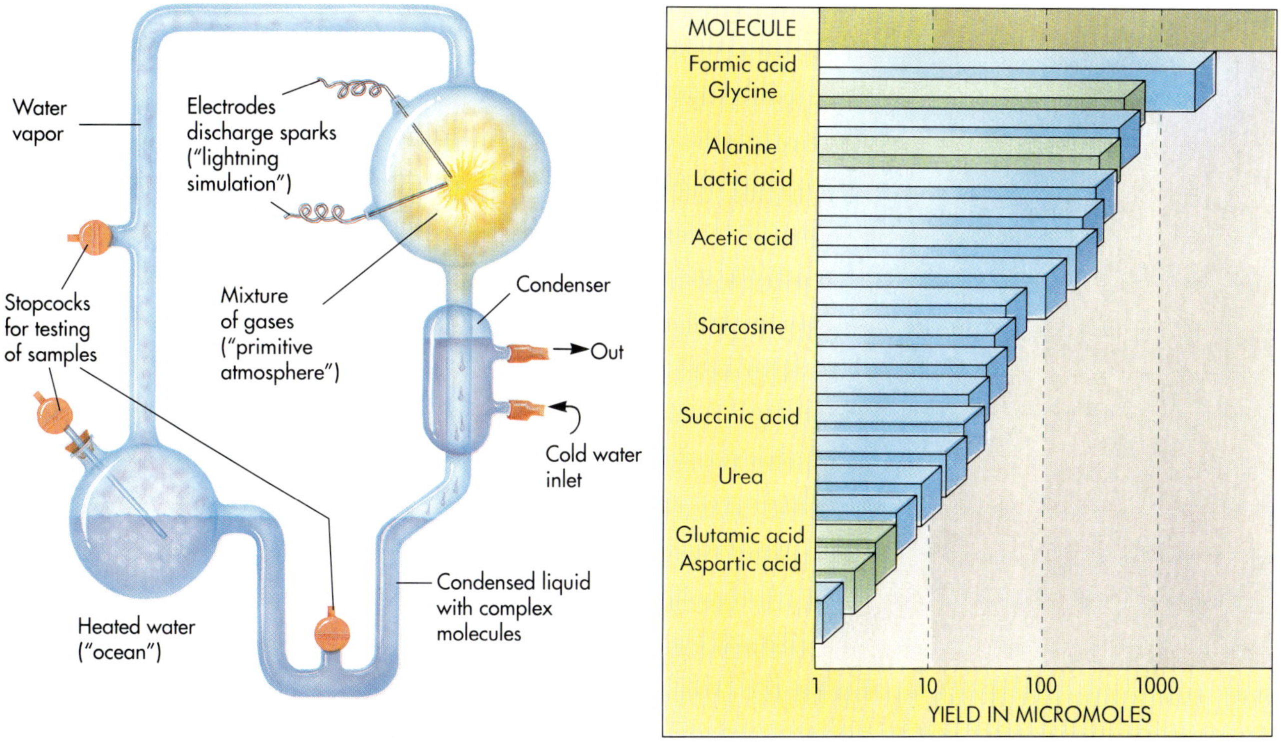

Miller-Urey Experiment
Figure 4–3

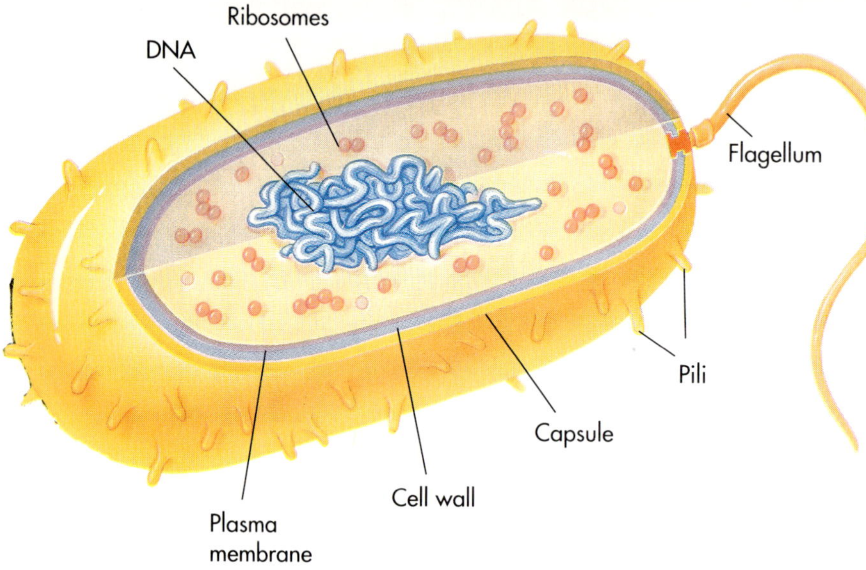

Structure of a Bacterial Cell
Figure 5–6

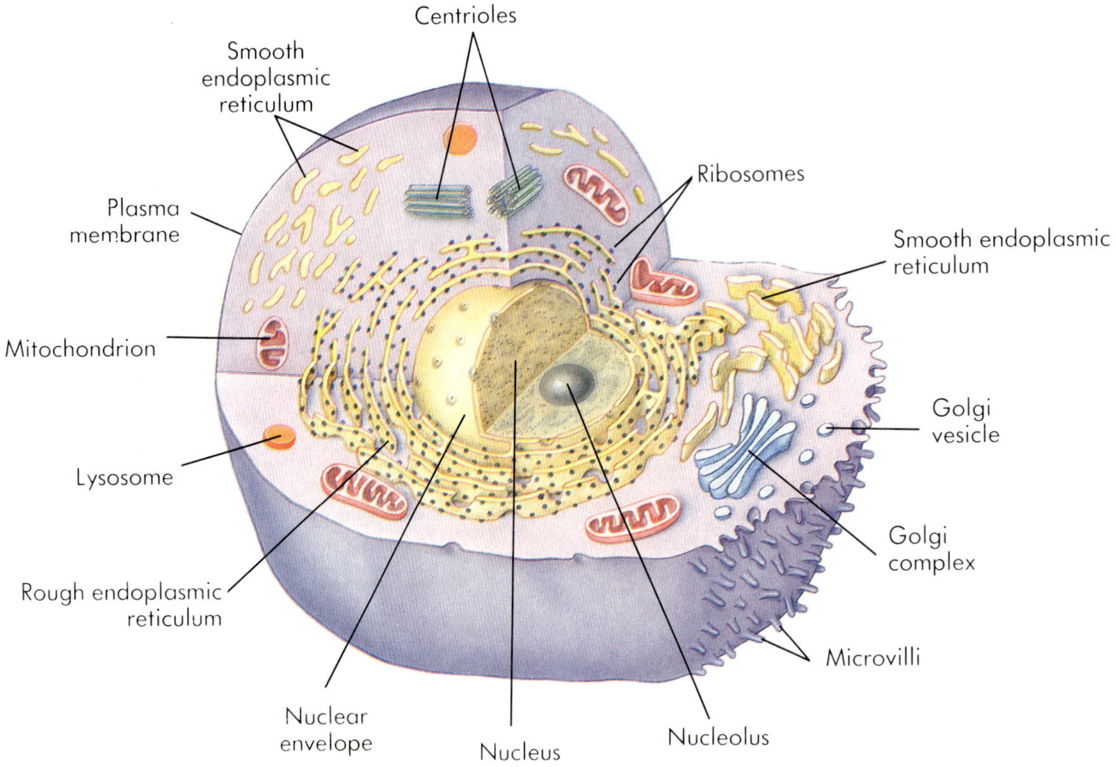

Animal Cell
Figure 5–8

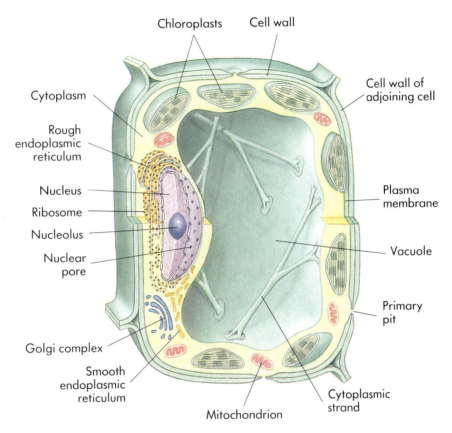

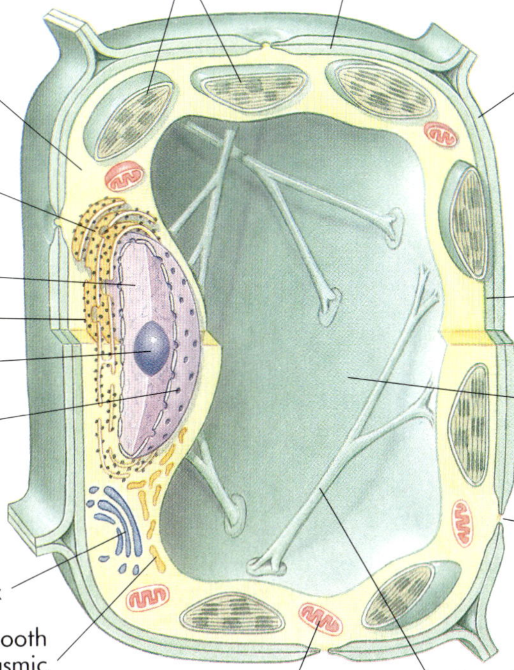

Plant Cell
Figure 5–9

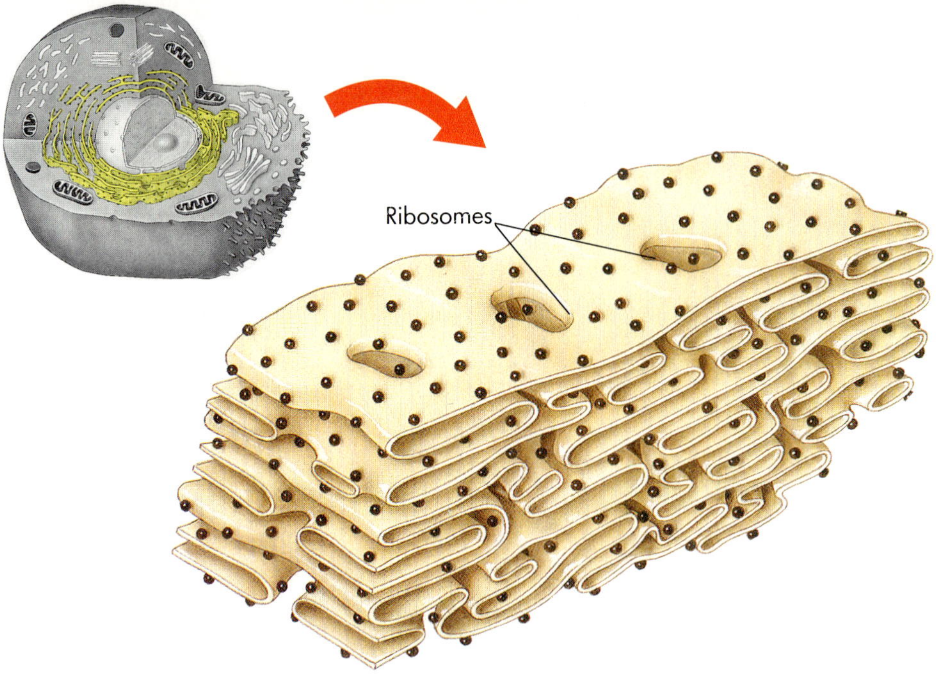

Rough Endoplasmic Reticulum
Figure 5–11

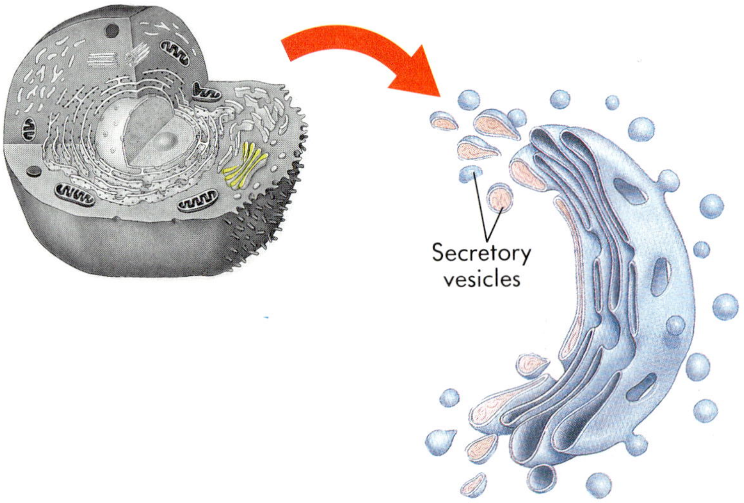

Golgi Body
Figure 5–17

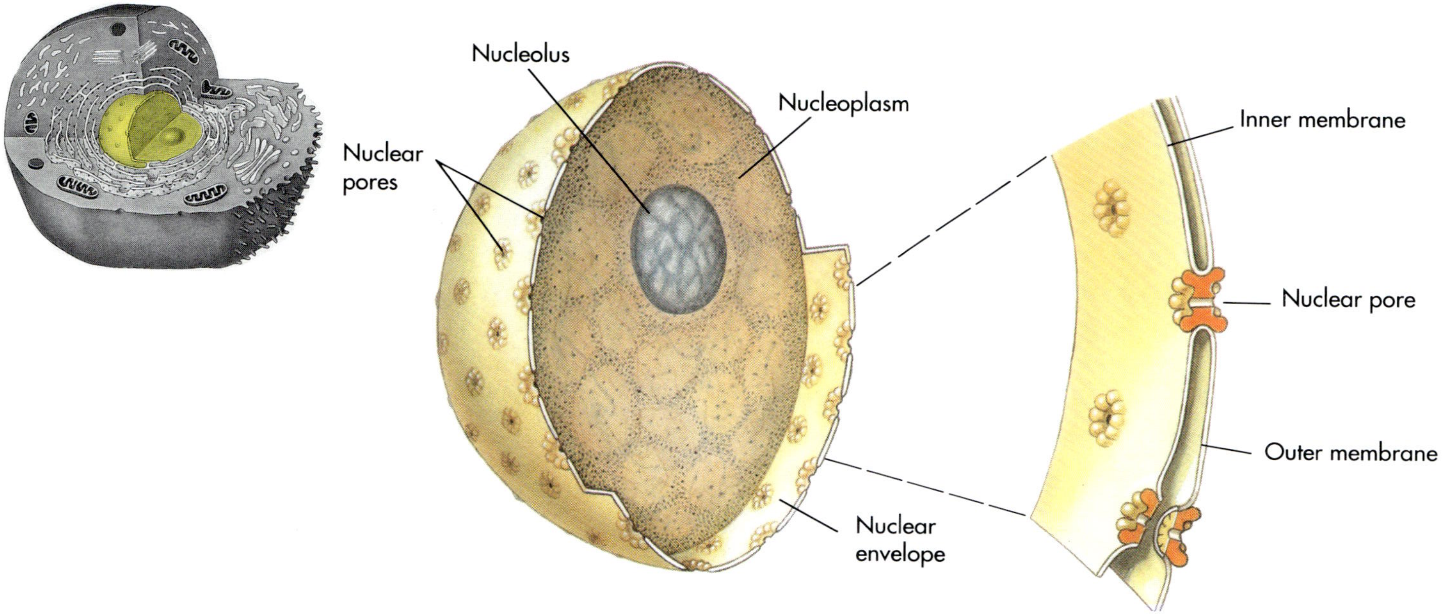

Nucleus
Figure 5–13

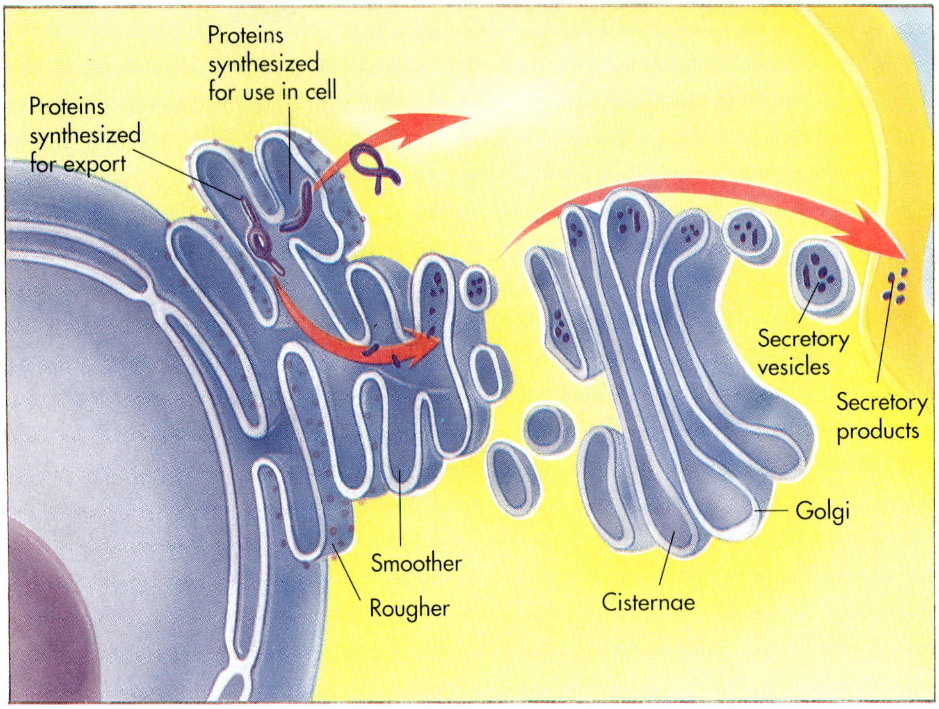

How Proteins Are Secreted Across Membranes
Figure 5–18

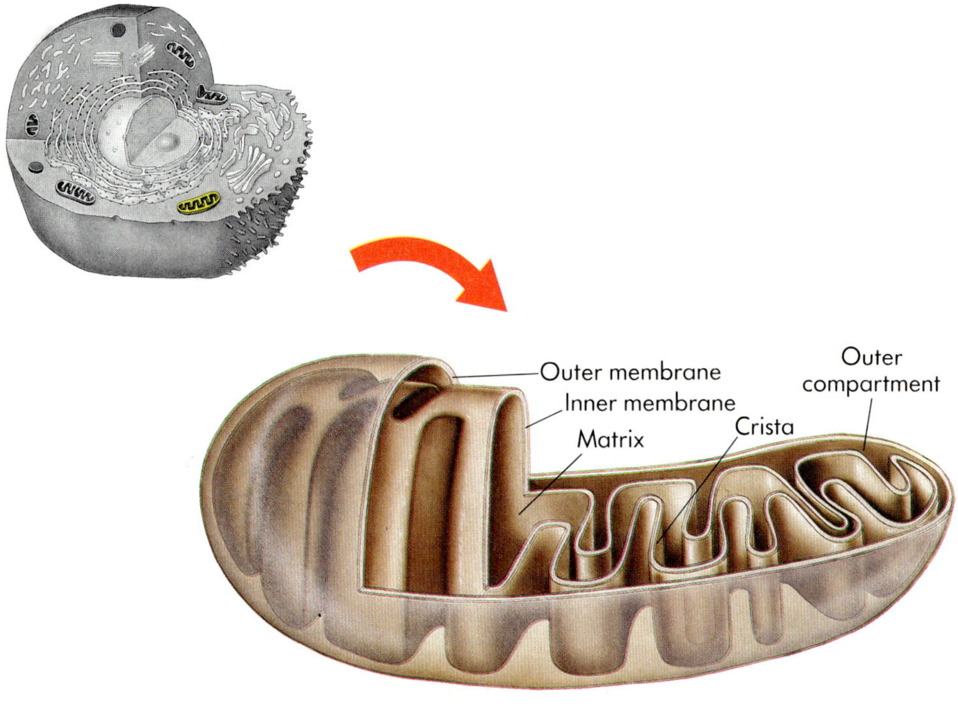

Mitochondria in Longitudinal and Cross Section
Figure 5–20

notes pg 40

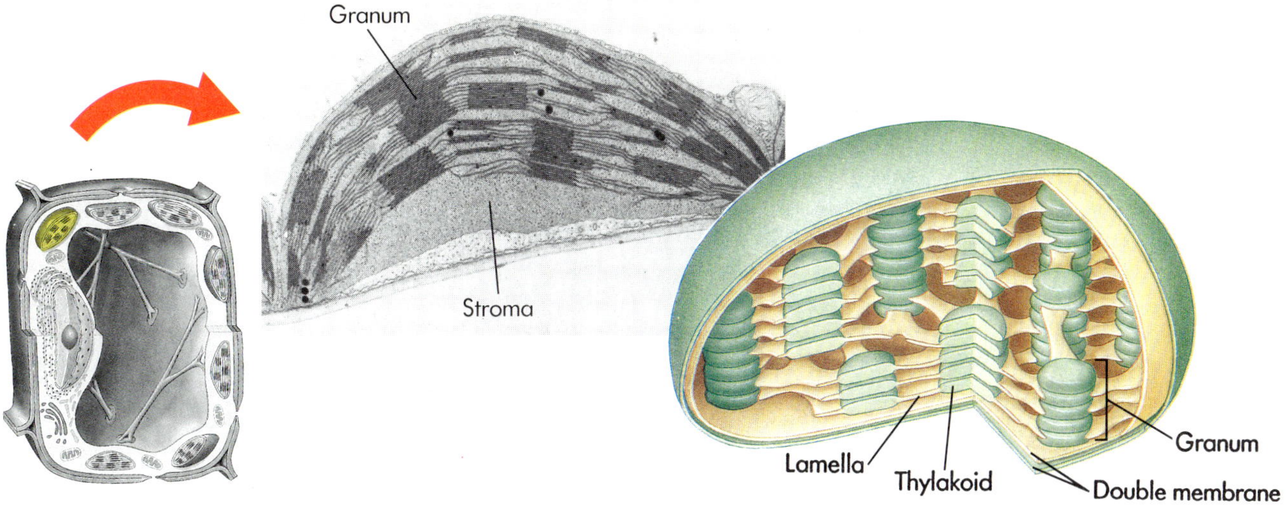

Chloroplast Structure
Figure 5–22

Photo by Kenneth Miller.

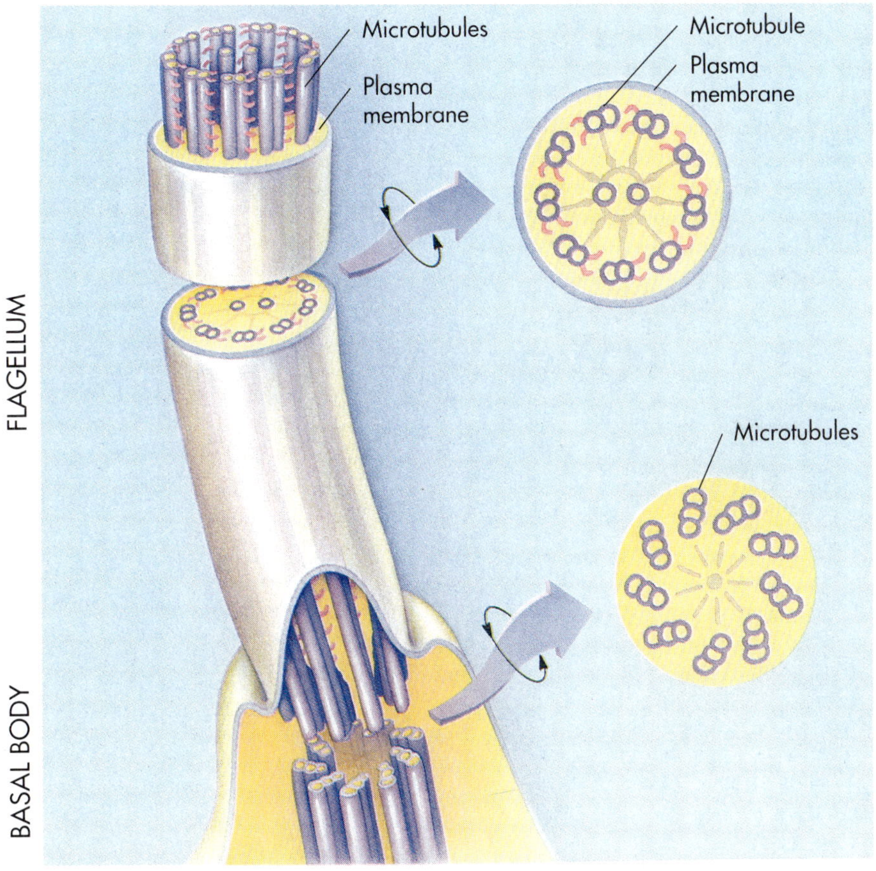

Structure of a Flagellum
Figure 5–28

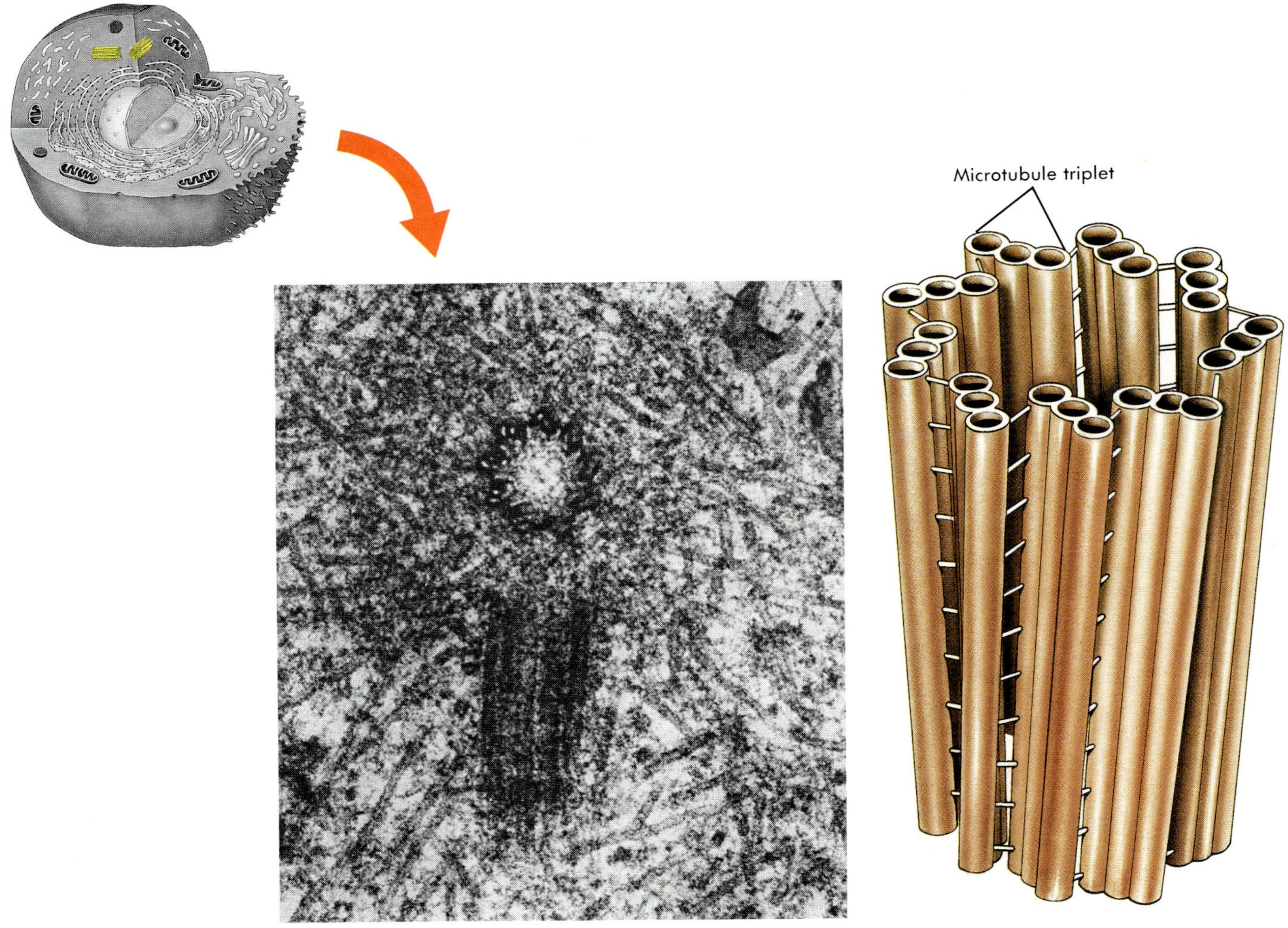

Centrioles
Figure 5–23

Photo by Kent McDonald.

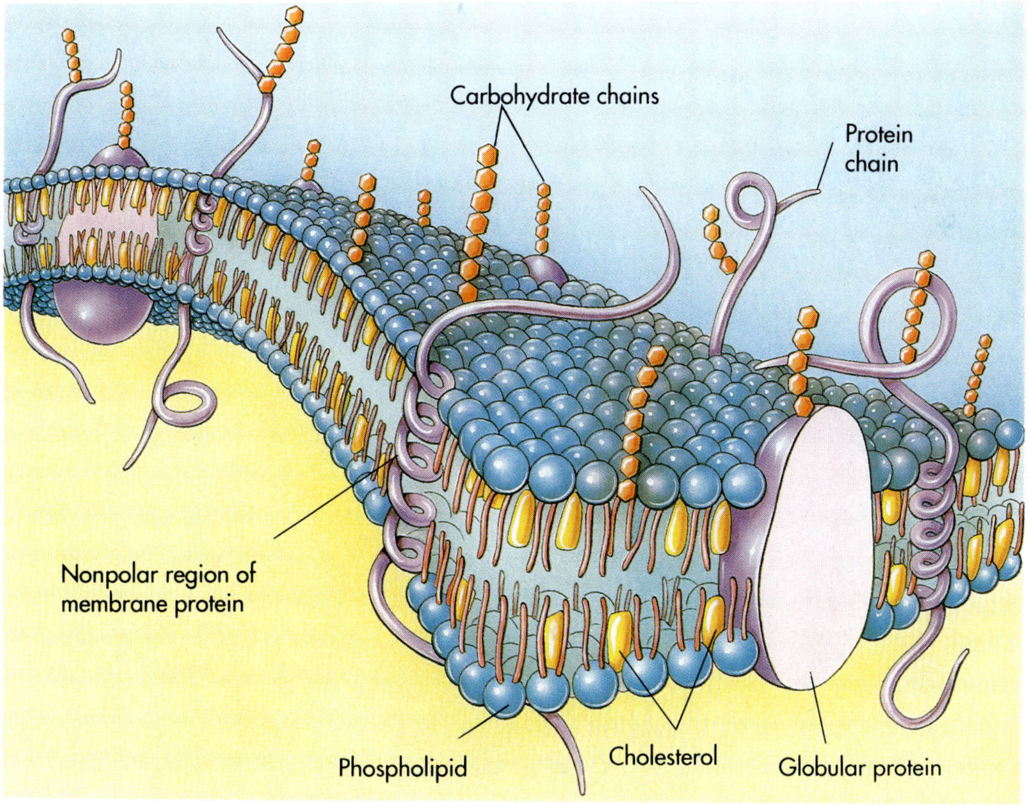

Cells Have Complex Surfaces
Figure 6–5

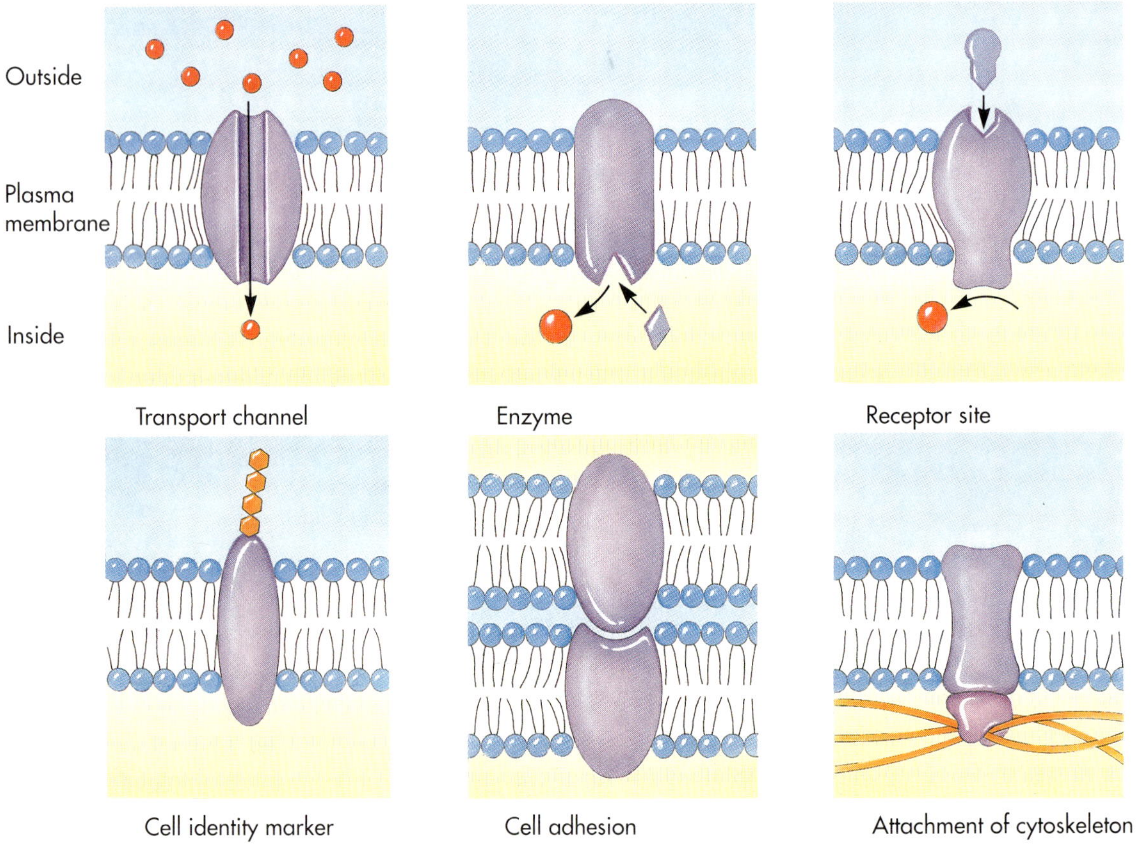

Functions of Plasma Membrane Proteins
Figure 6–8

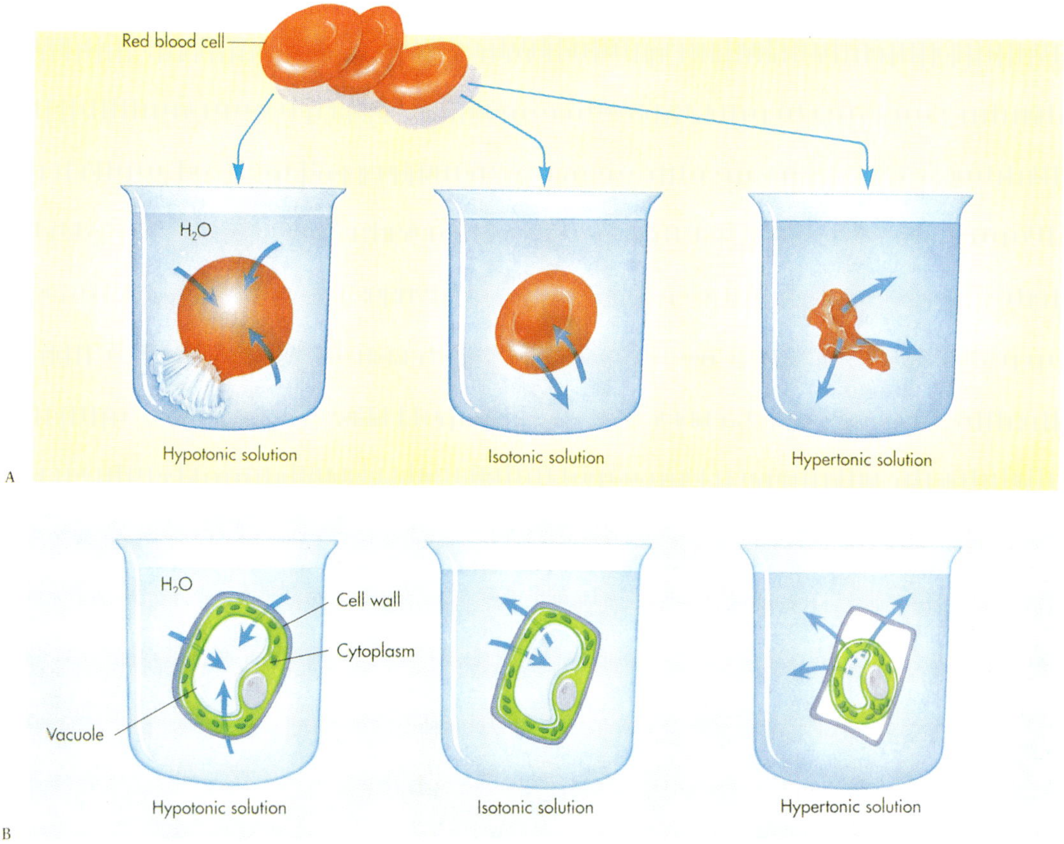

Osmosis
Figure 6–11

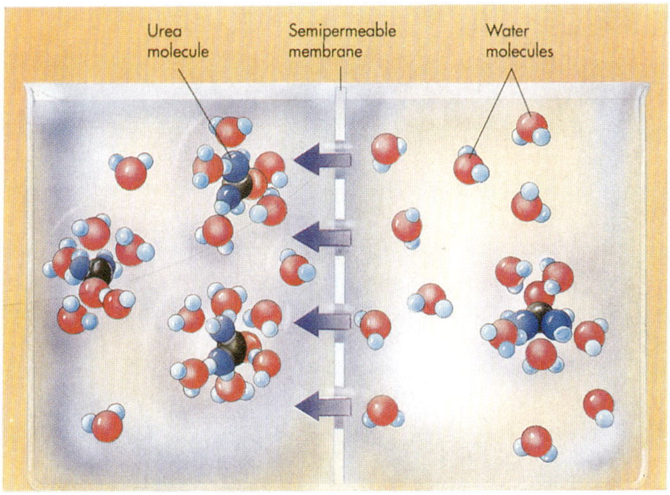

How Solutes Create Osmotic Pressure
Figure 6–12

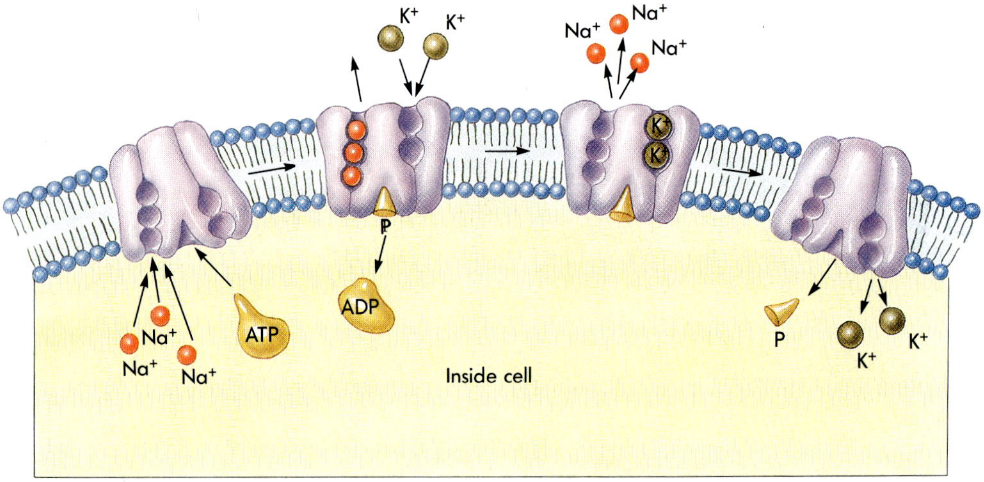

Sodium-Potassium Pump
Figure 6–20

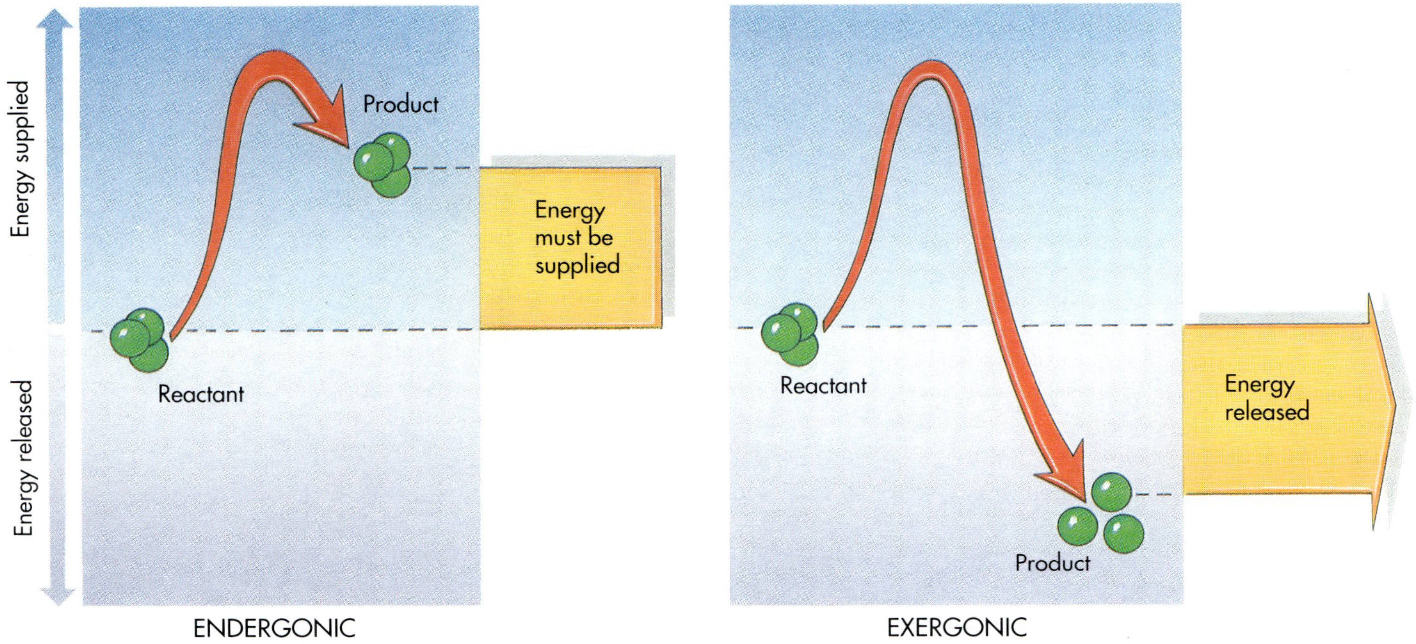

Energy in Chemical Reactions
Figure 7–7

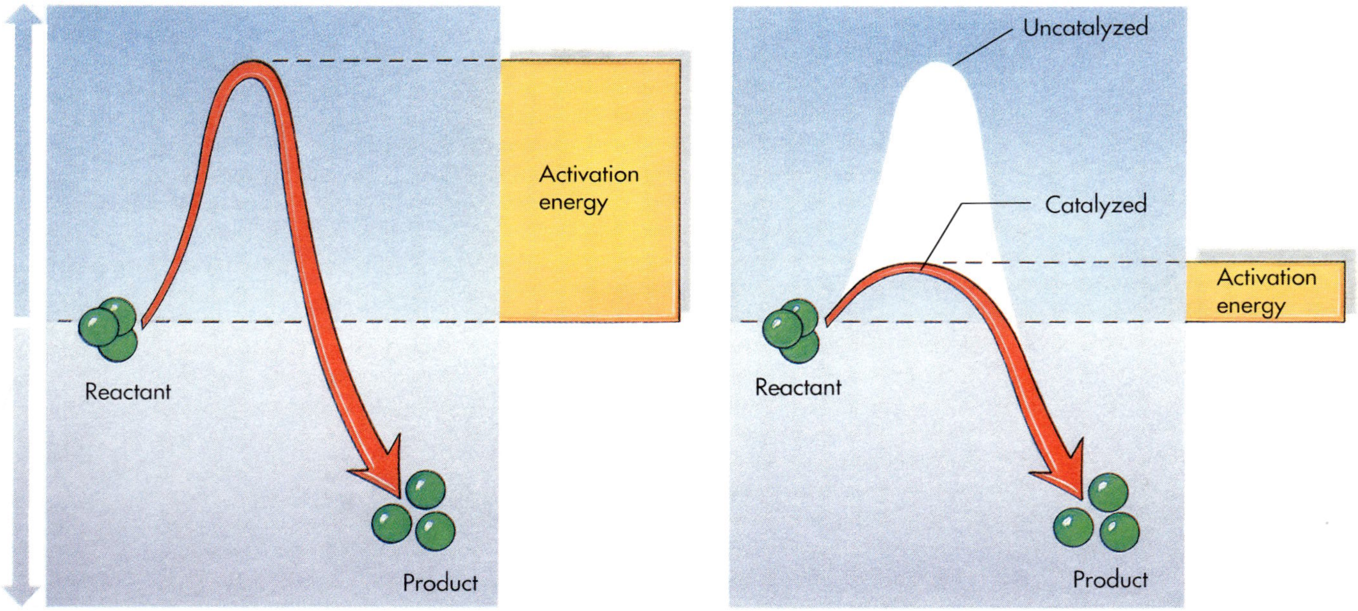

Exergonic Reactions
Figure 7–8

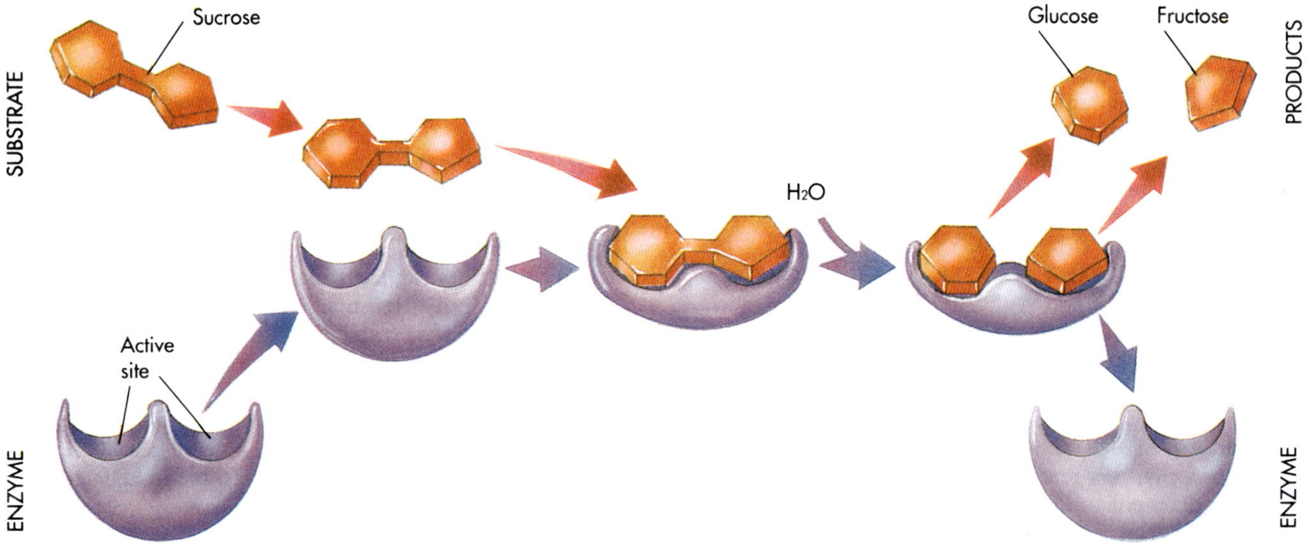

Catalytic Cycle of an Enzyme
Figure 7–10

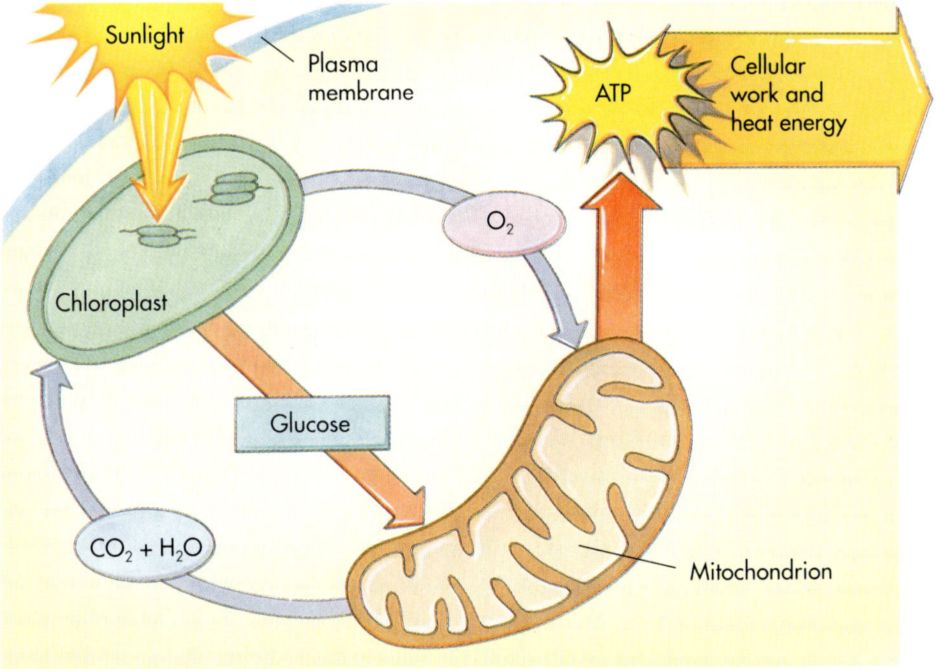

Overview of Aerobic Metabolism
Figure 7–18

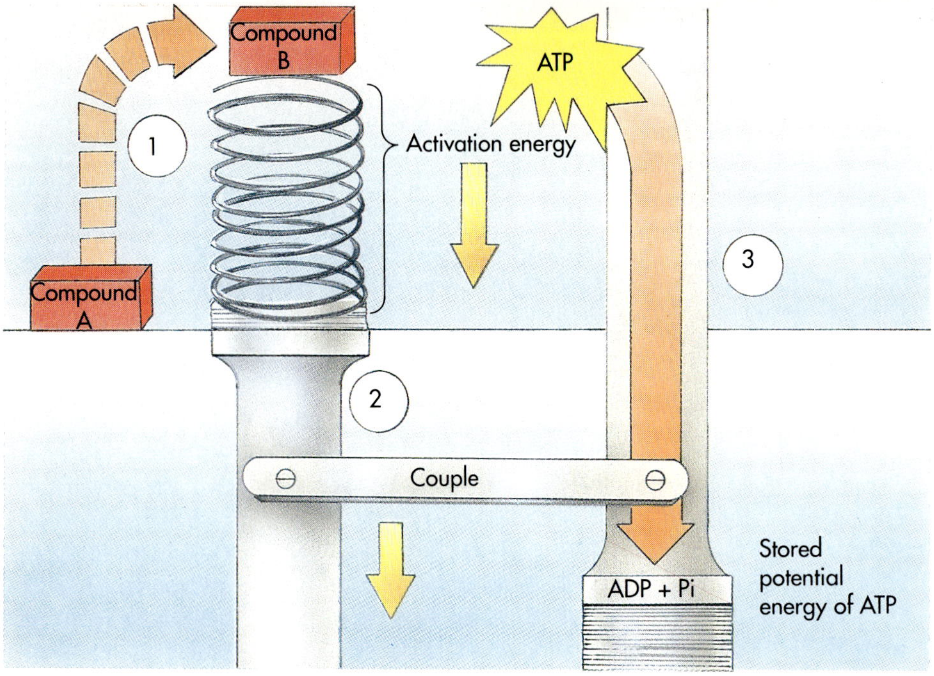

How Coupled Reactions Work
Figure 8–1

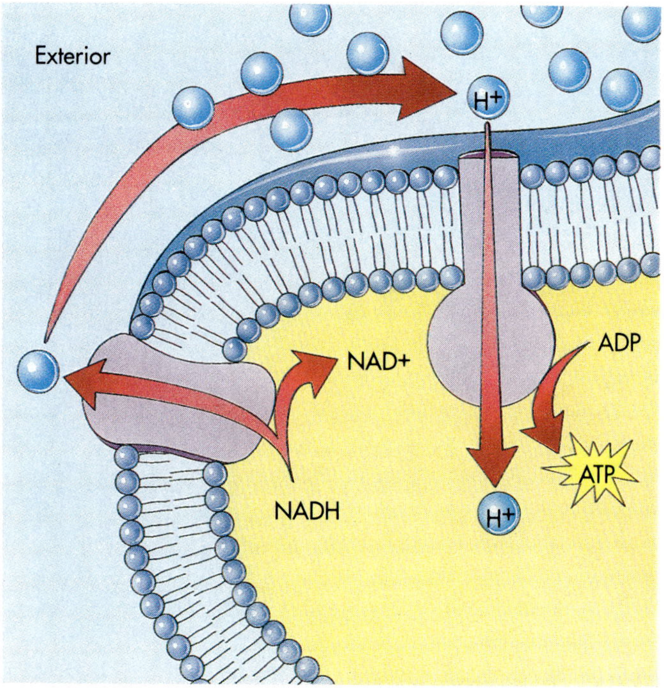

Chemiosmosis
Figure 8–3

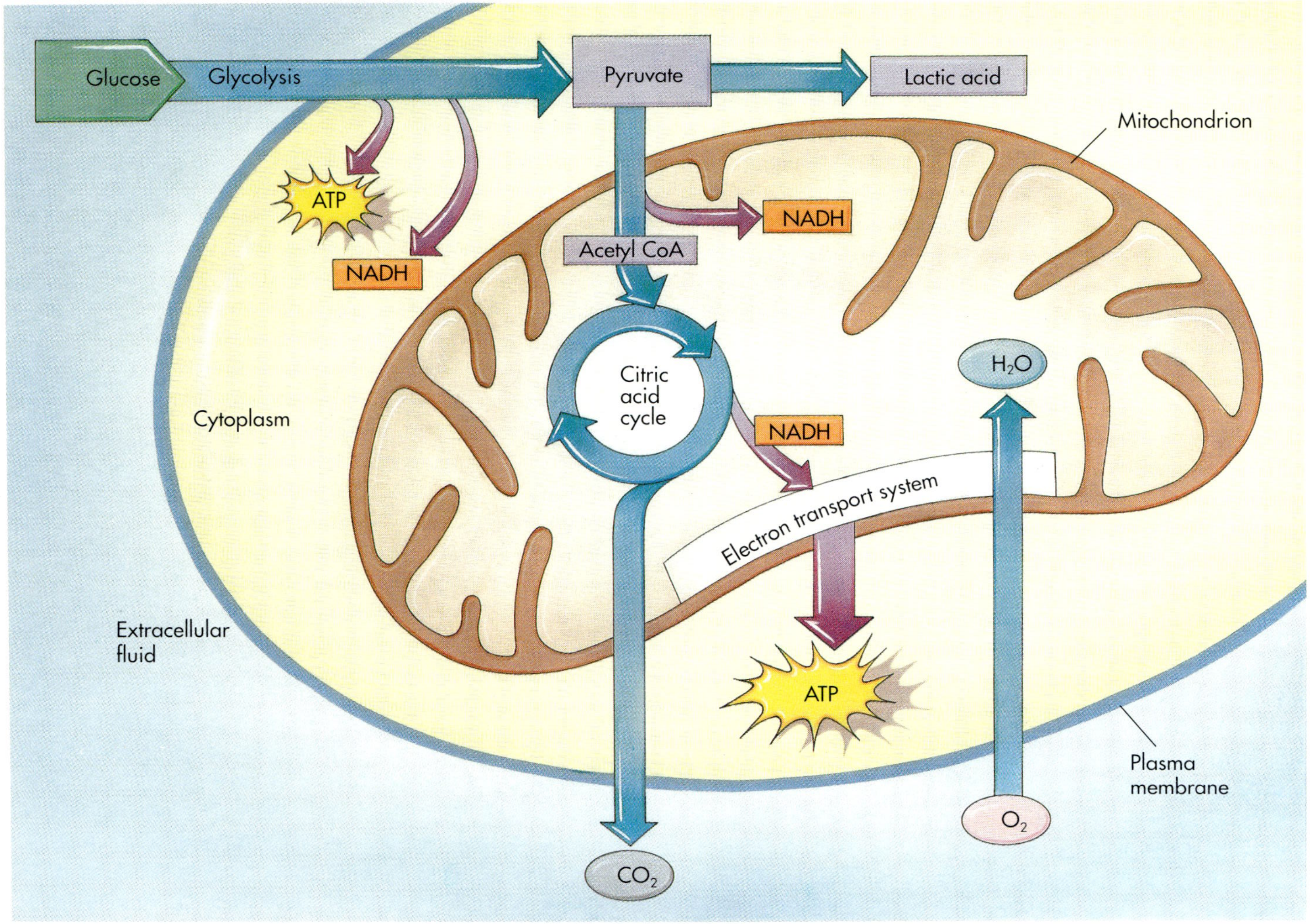

Overview of Cellular Respiration
Figure 8–4

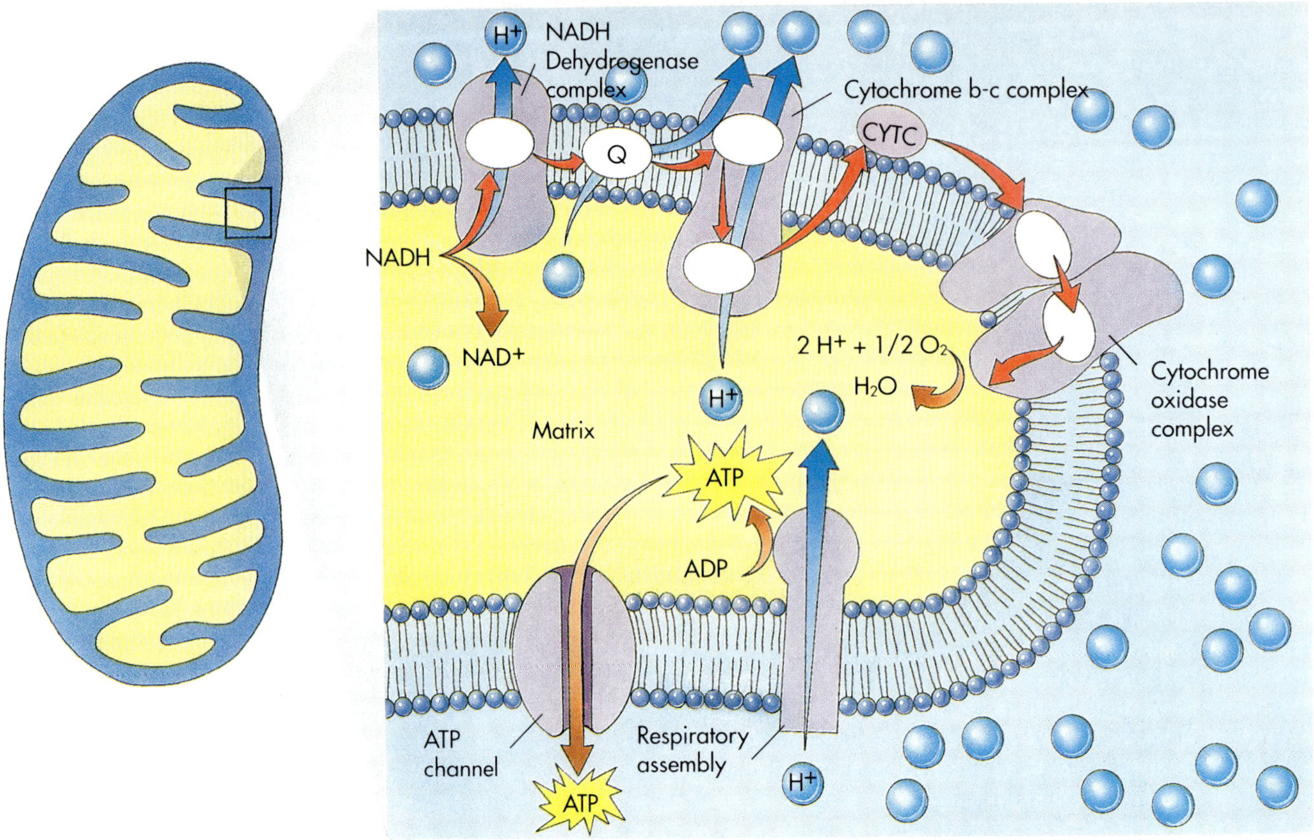

Electron Transport Chain
Figure 8–14

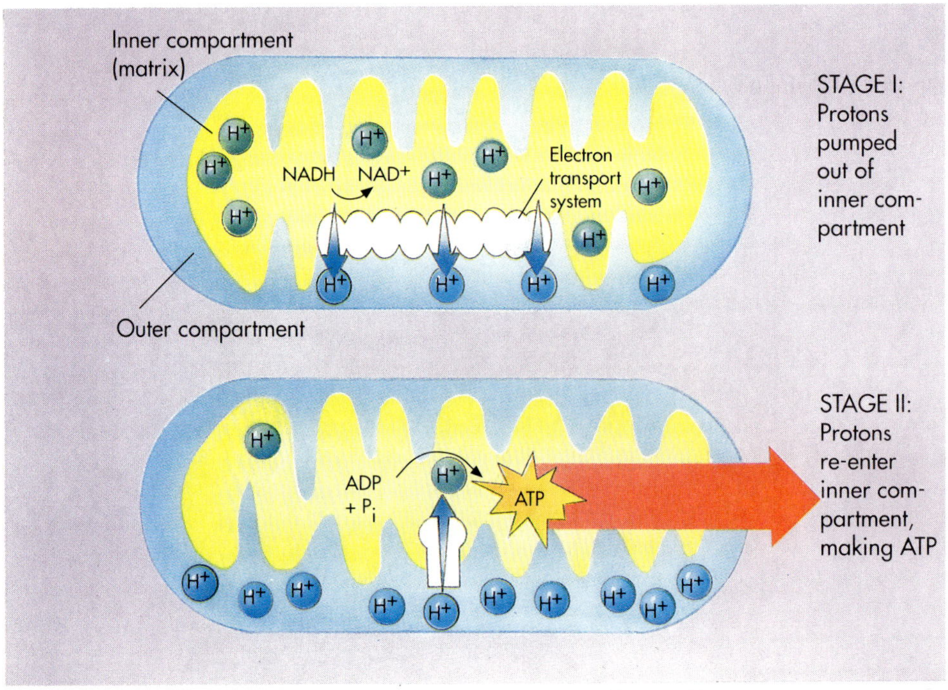

Proton Movement in the Mitochondrion
Figure 8–15

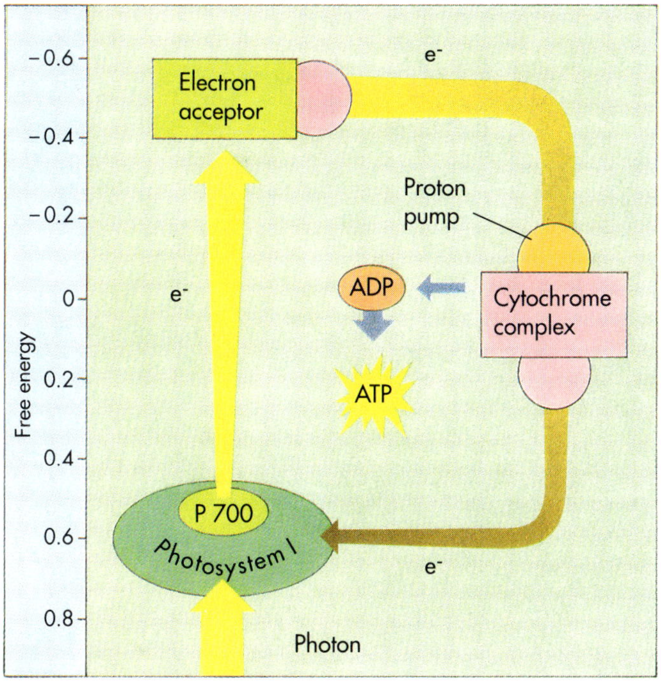

Path of an Electron in Bacterial Photosynthesis
Figure 9–7

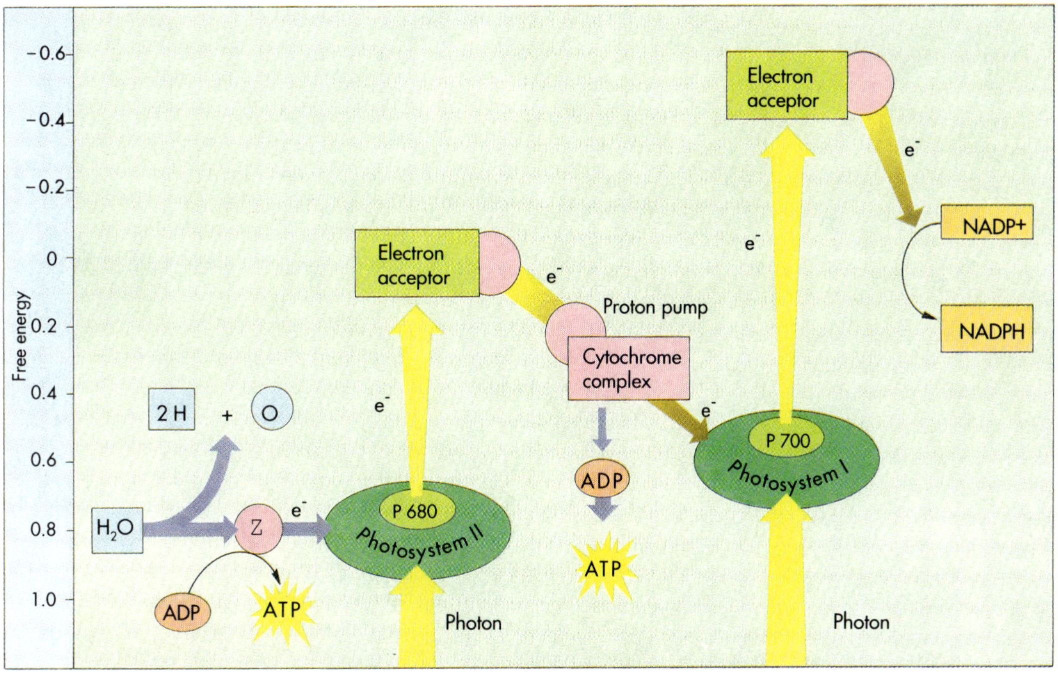

Path of an Electron in a Chloroplast
Figure 9–8

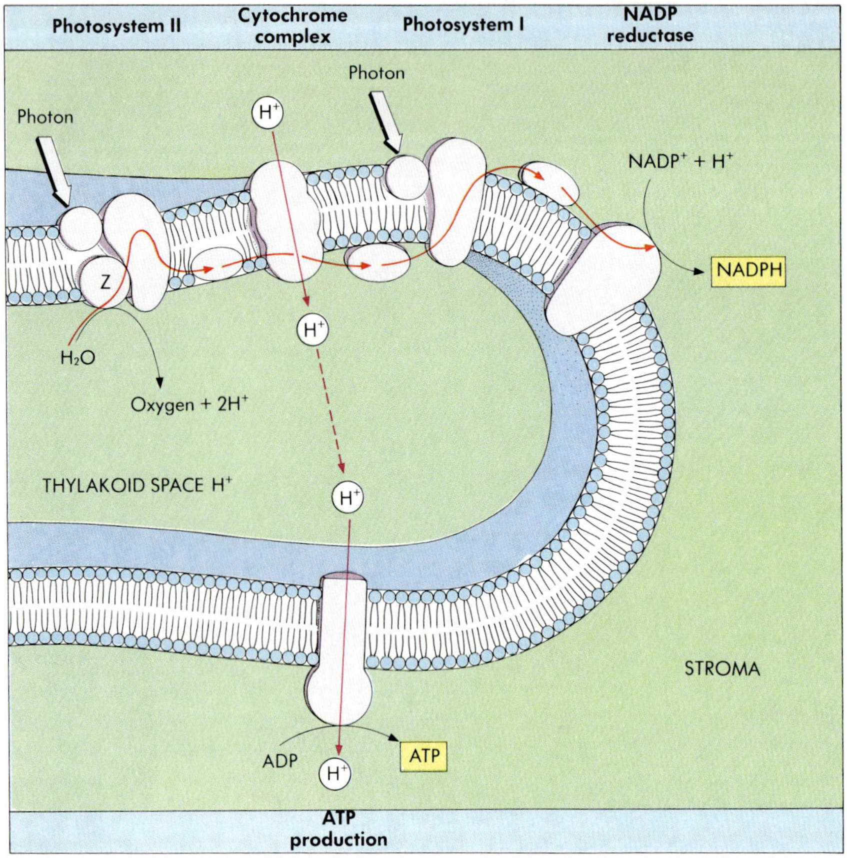

Photosynthetic Electron Transport System
Figure 9–9

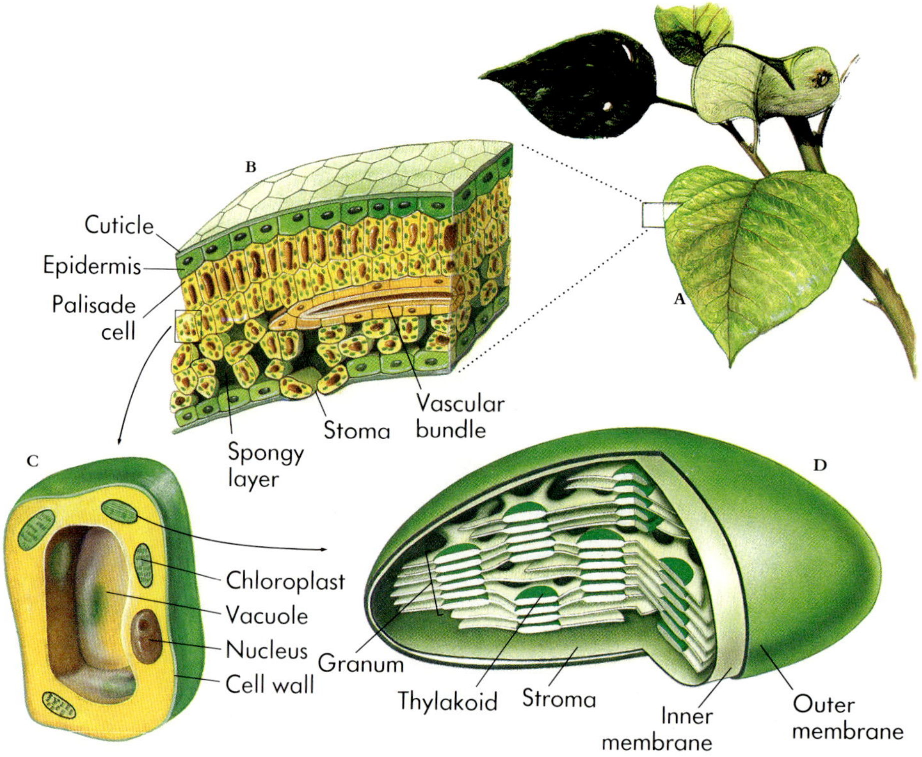

Journey into a Leaf
Figure 9–14

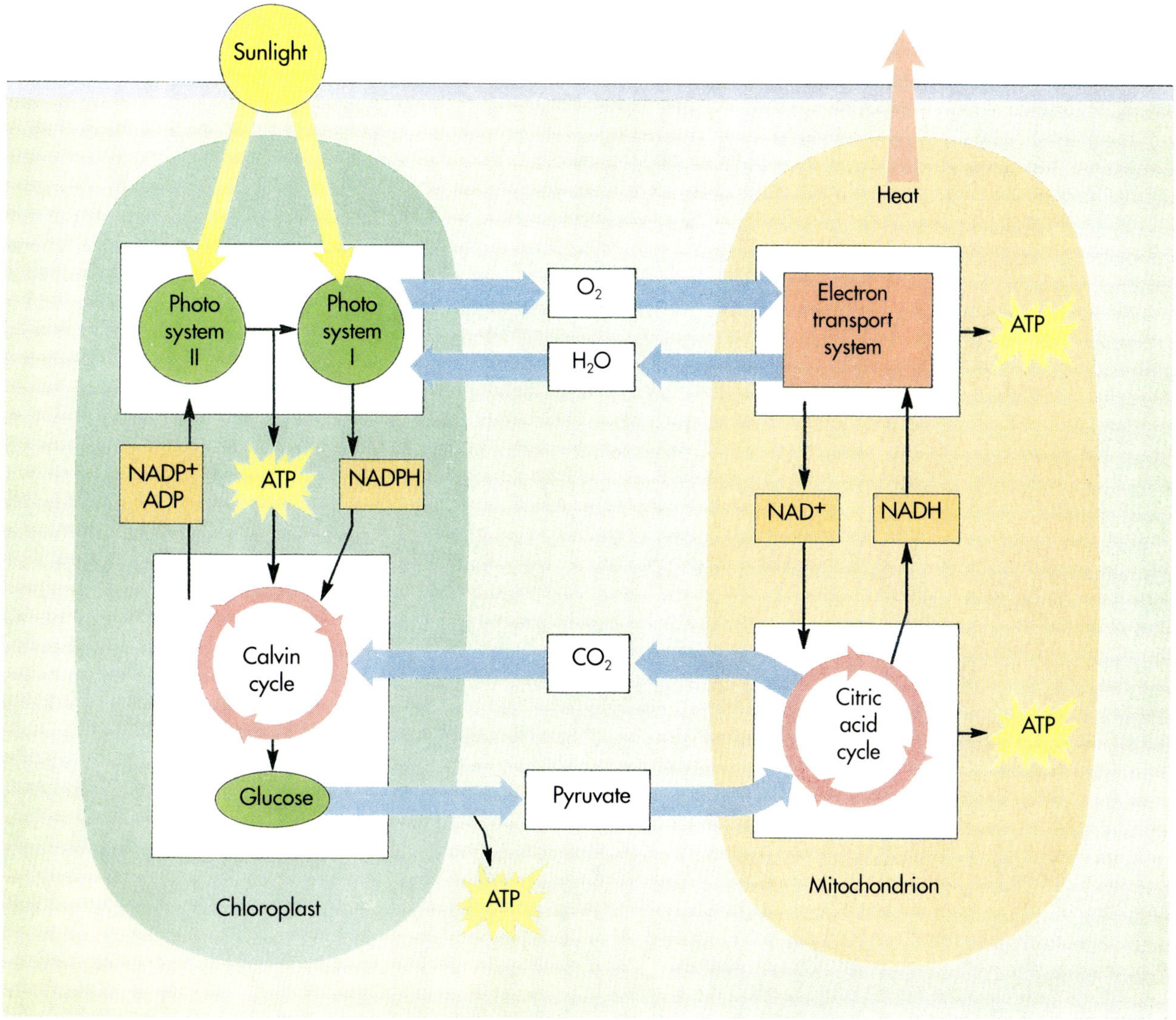

Metabolic Machine
Figure 9–21

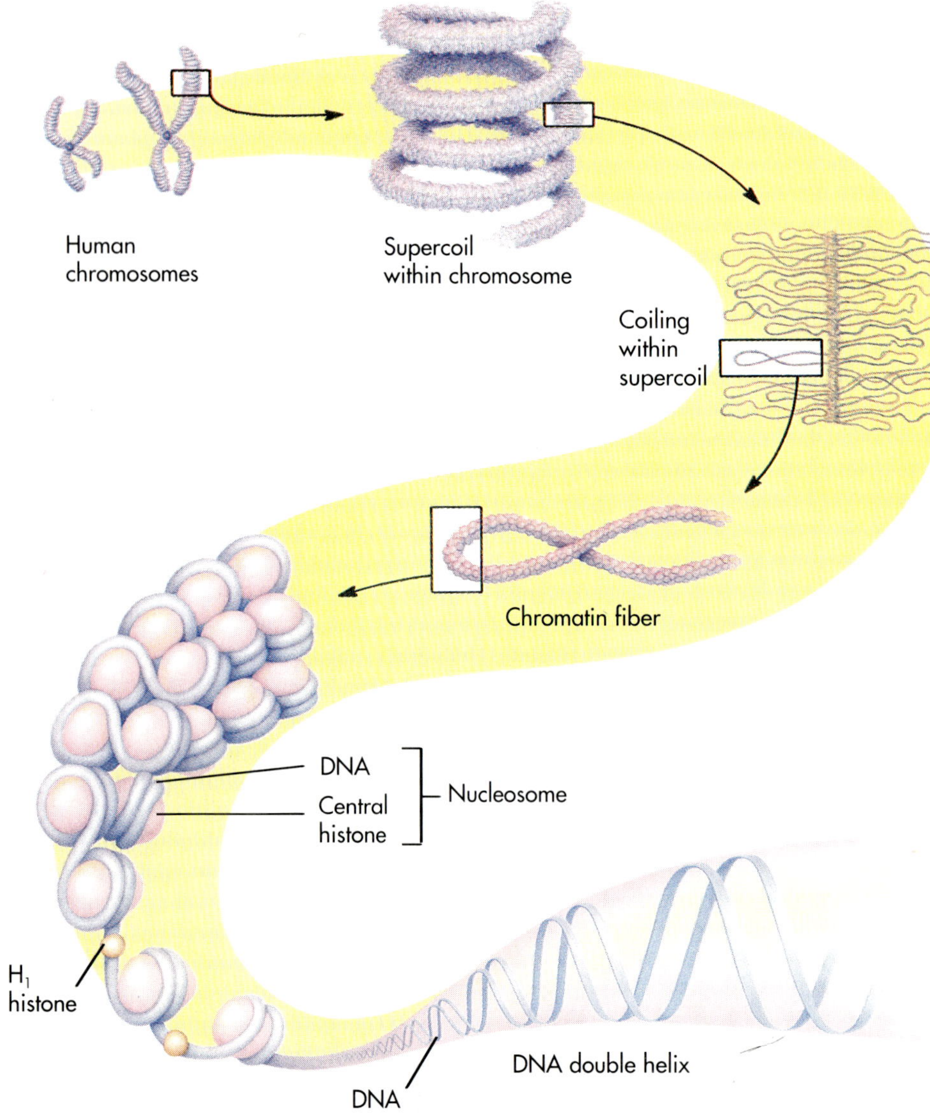

Levels of Chromosomal Organization
Figure 10–4

Sexual Life Cycle
Figure 11–4

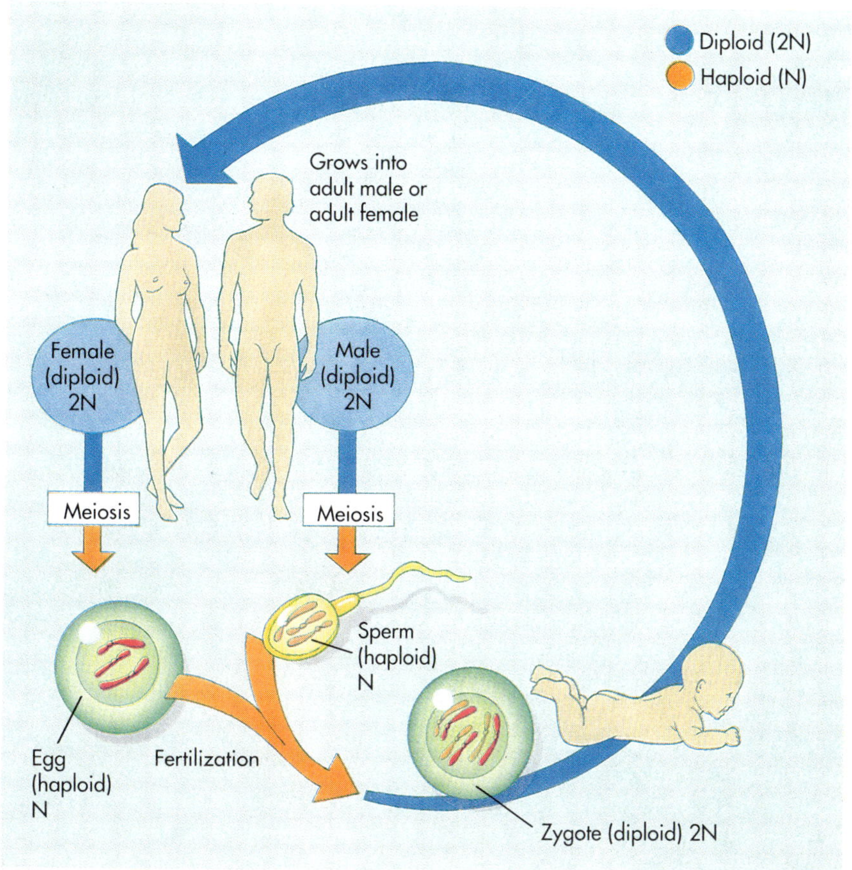

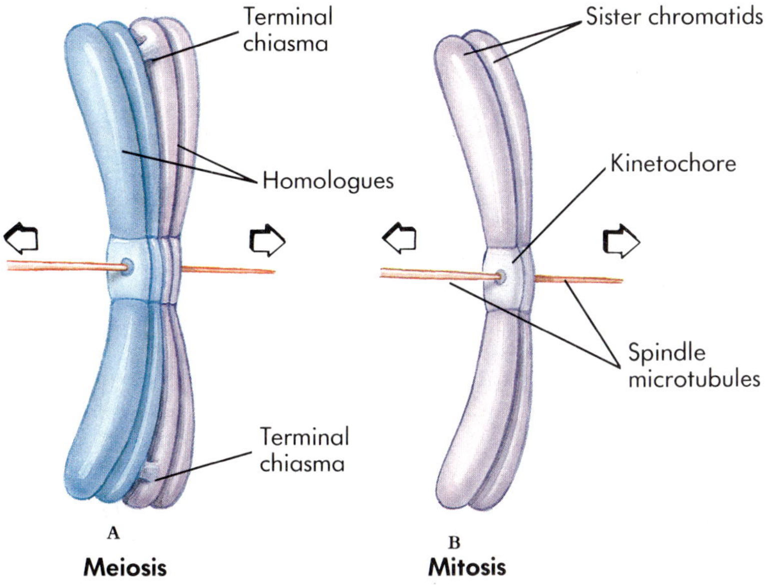

Presence of Chiasmata in Reduction Division in Meiosis
Figure 11–7

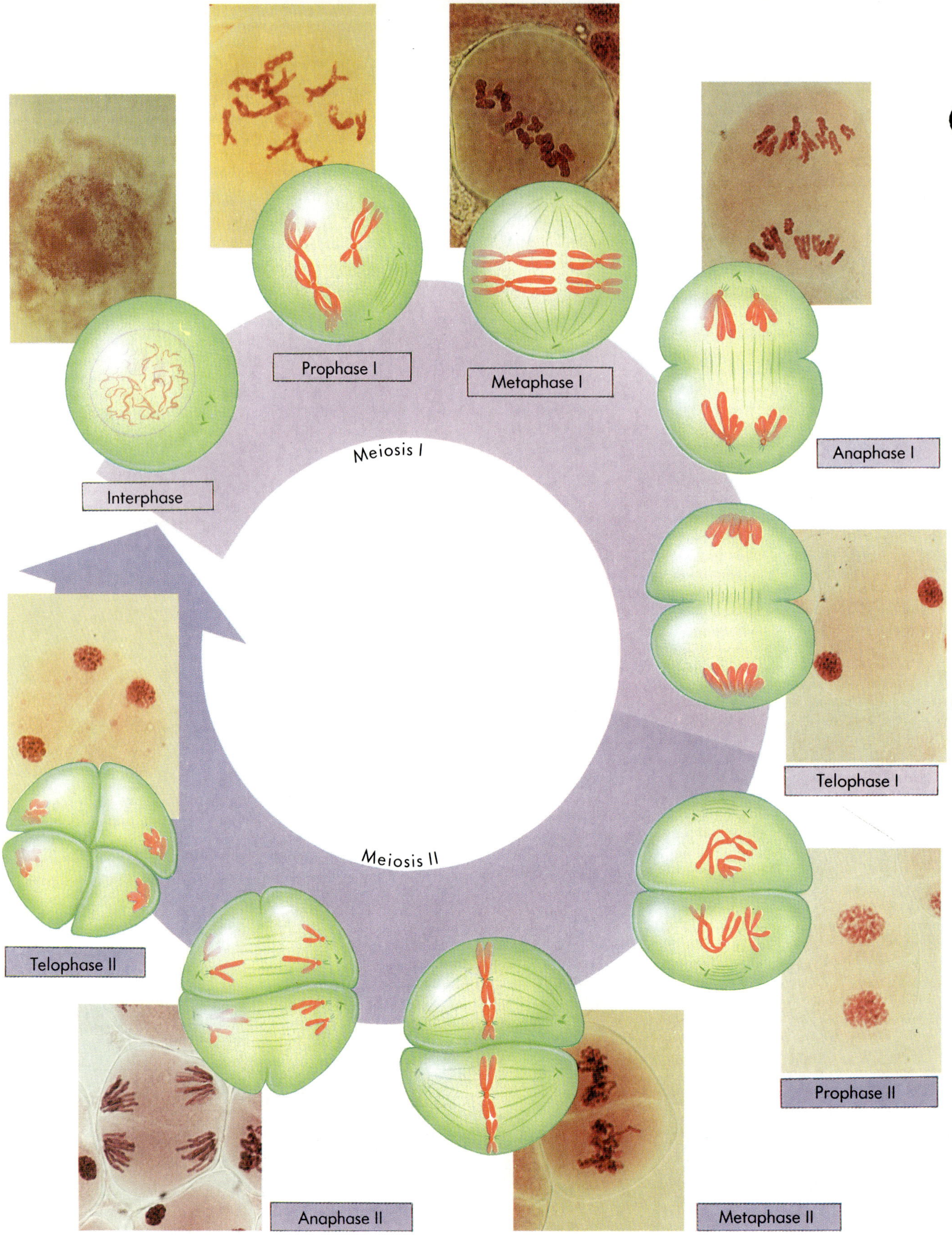

Meiosis
Figure 11–9

Photos by C. A. Hasenkampf/BPS.

MITOSIS

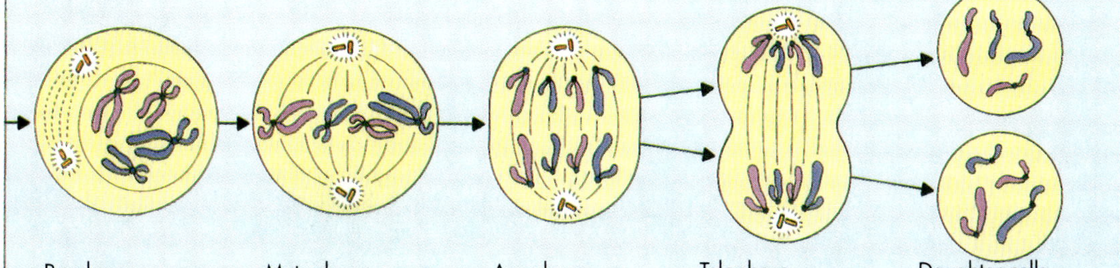

Prophase
Chromosomes condense. Spindle fibers form between centrioles which move toward opposite poles.

Metaphase
Microtubule spindle apparatus attaches to chromosomes. Chromosomes align along spindle equator.

Anaphase
Sister chromatids separate and move to opposite poles.

Telophase
Chromatids arrive at each pole, and new nuclear membranes form. Cell division begins.

Daughter cells
Cell division complete. Each cell receives chromosomes which are identical to those in original nucleus.

MEIOSIS I

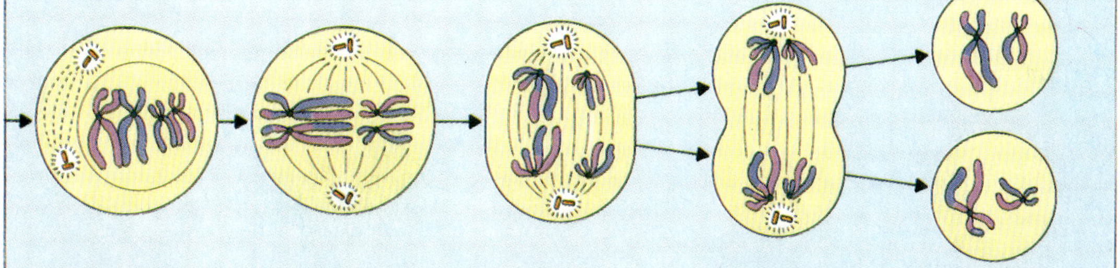

Prophase I
Homologous chromosomes further condense and pair. Crossing over occurs. Spindle fibers form between centrioles which move toward opposite poles.

Metaphase I
Microtubule spindle apparatus attaches to chromosomes. Homologous pairs align along spindle equator.

Anaphase I
Homologous pairs of chromosomes separate and move to opposite poles.

Telophase I
One set of paired chromosomes arrives at each pole, and nuclear division begins.

Daughter cells
Each cell receives exchanged chromosomal material from homologous chromosomes.

MEIOSIS II

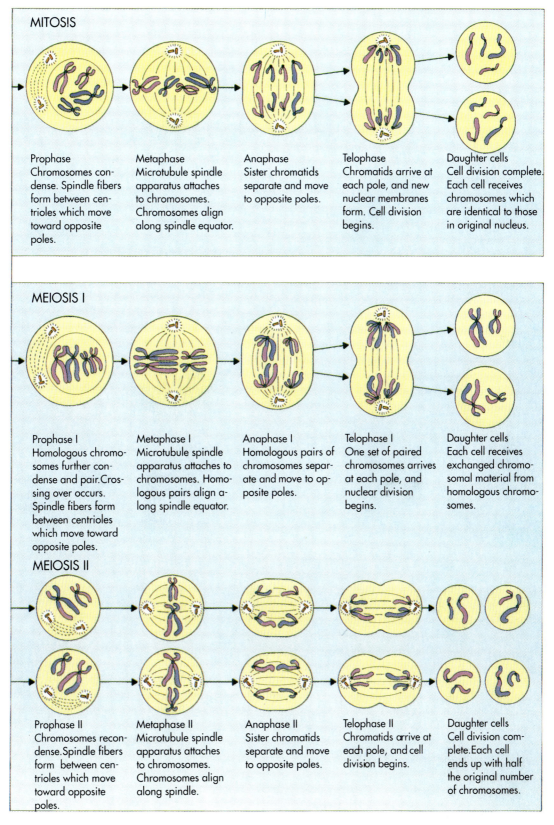

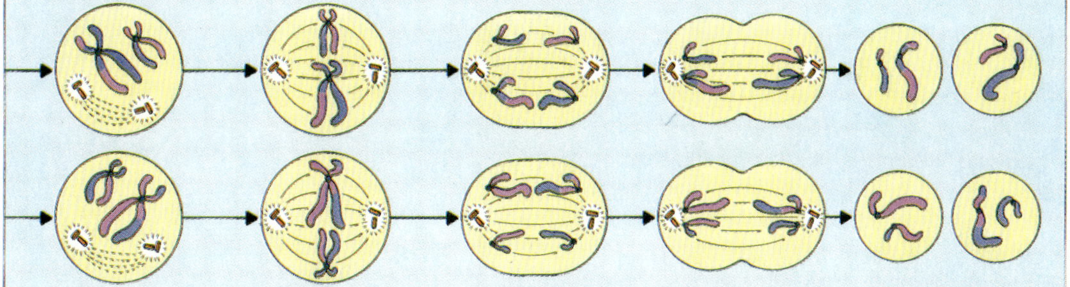

Prophase II
Chromosomes recondense. Spindle fibers form between centrioles which move toward opposite poles.

Metaphase II
Microtubule spindle apparatus attaches to chromosomes. Chromosomes align along spindle.

Anaphase II
Sister chromatids separate and move to opposite poles.

Telophase II
Chromatids arrive at each pole, and cell division begins.

Daughter cells
Cell division complete. Each cell ends up with half the original number of chromosomes.

Comparison of Meiosis and Mitosis
Figure 11–11

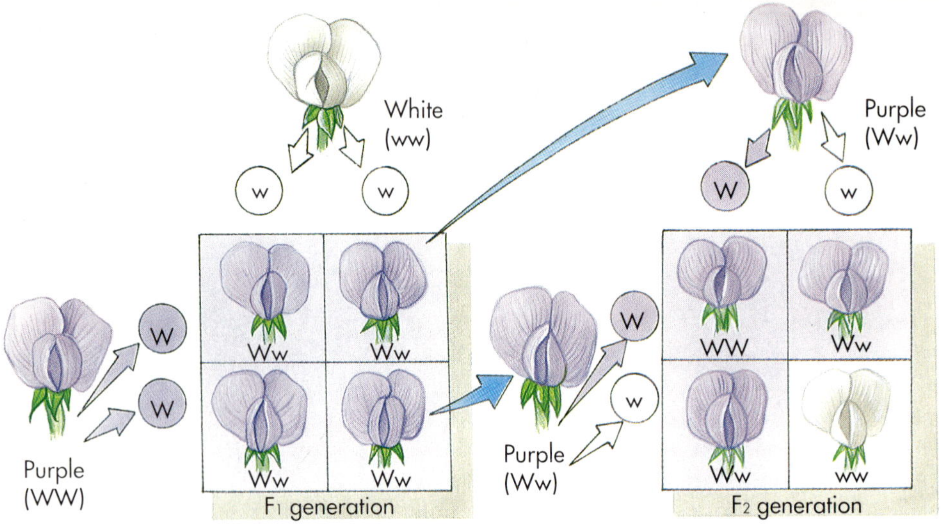

Mendel's Cross of Peas Differing in Flower Color
Figure 12–13

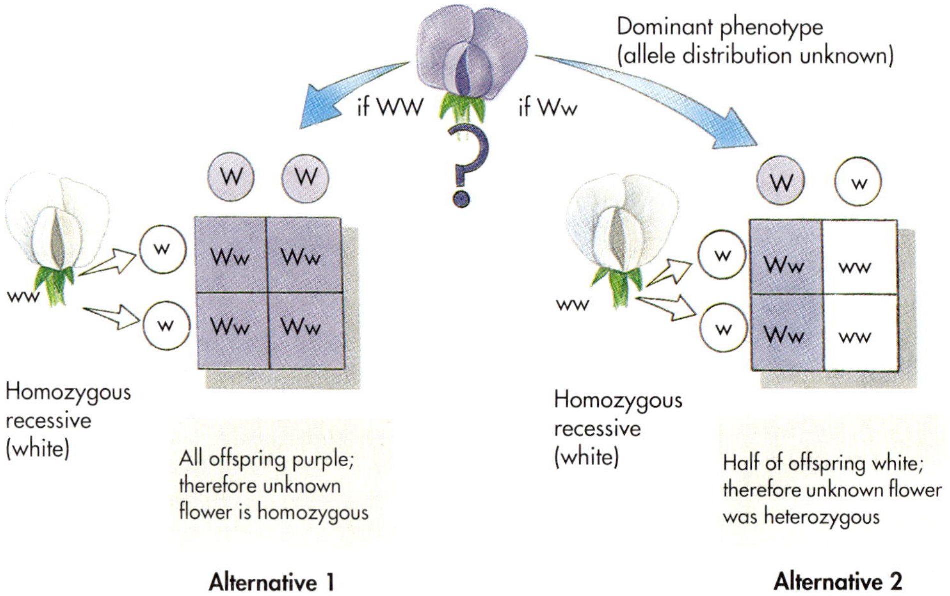

Testcross
Figure 12–15

Incomplete Dominance
Figure 12–24

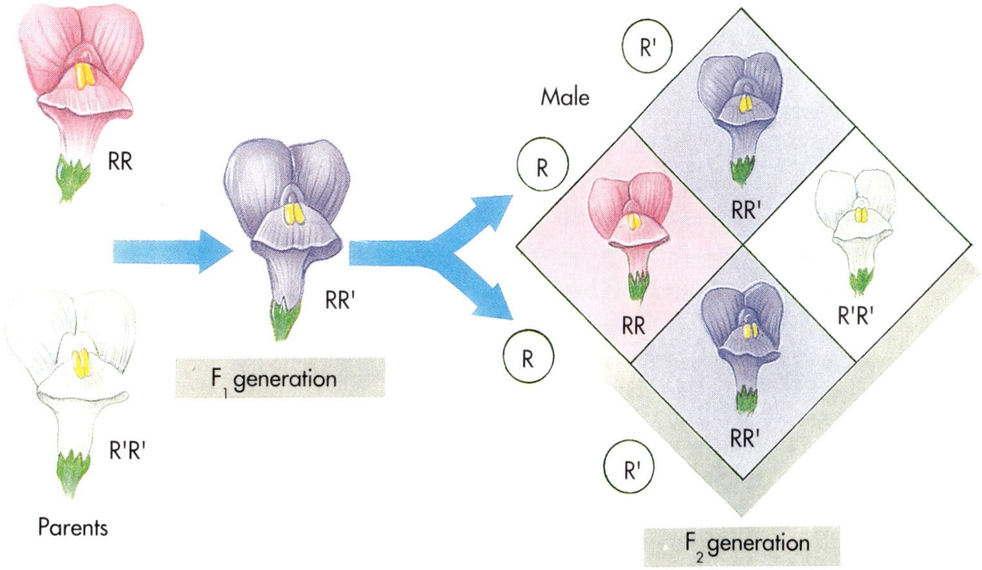

Amniocentesis
Figure 13–14

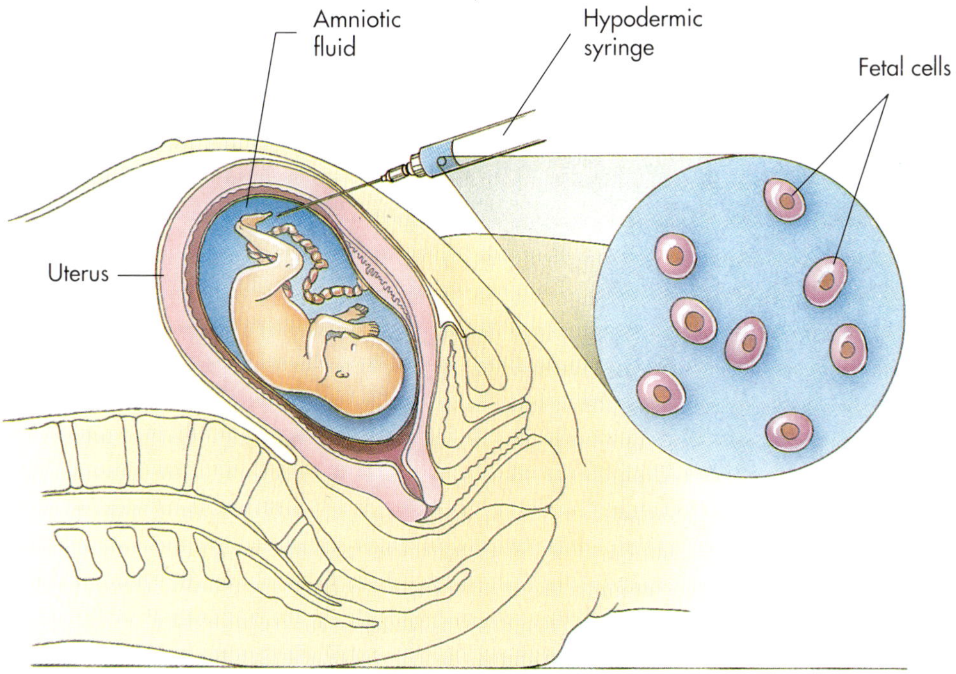

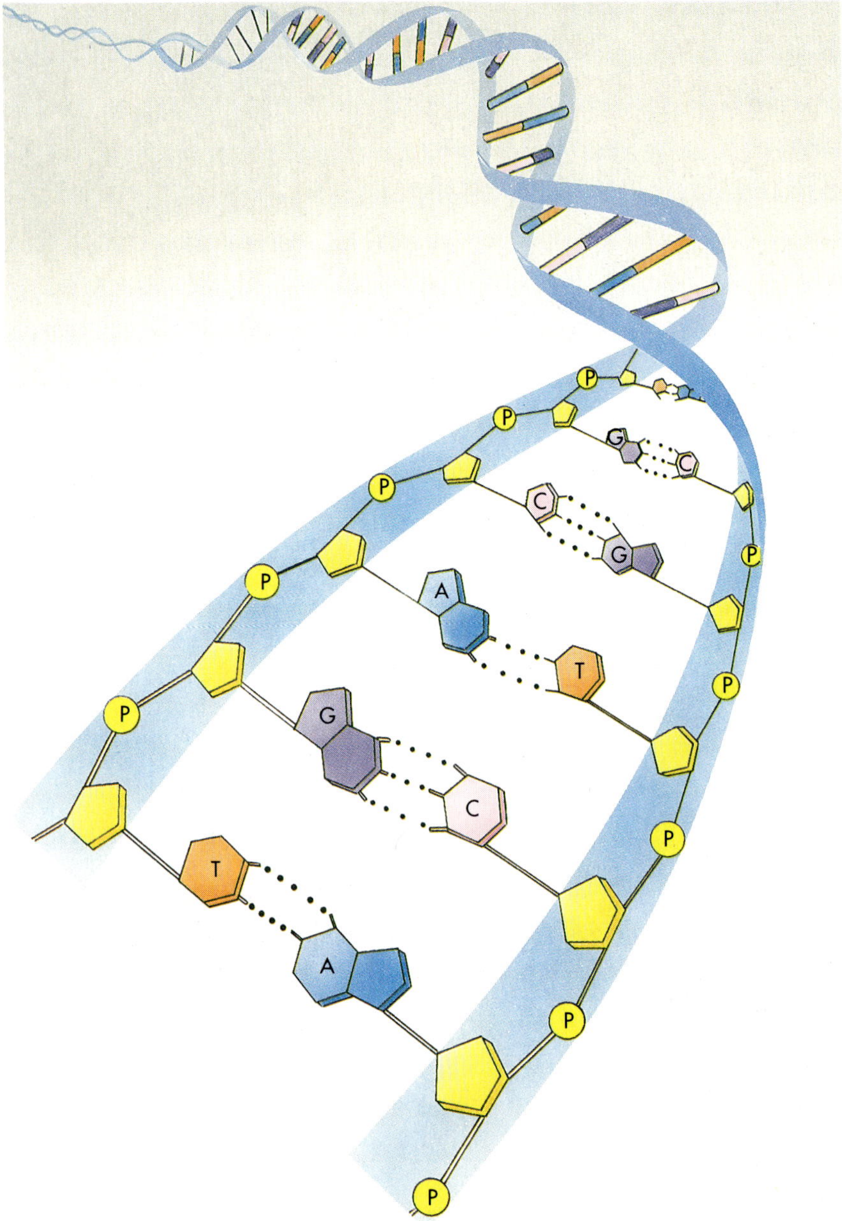

Base Pairing
Figure 14–16

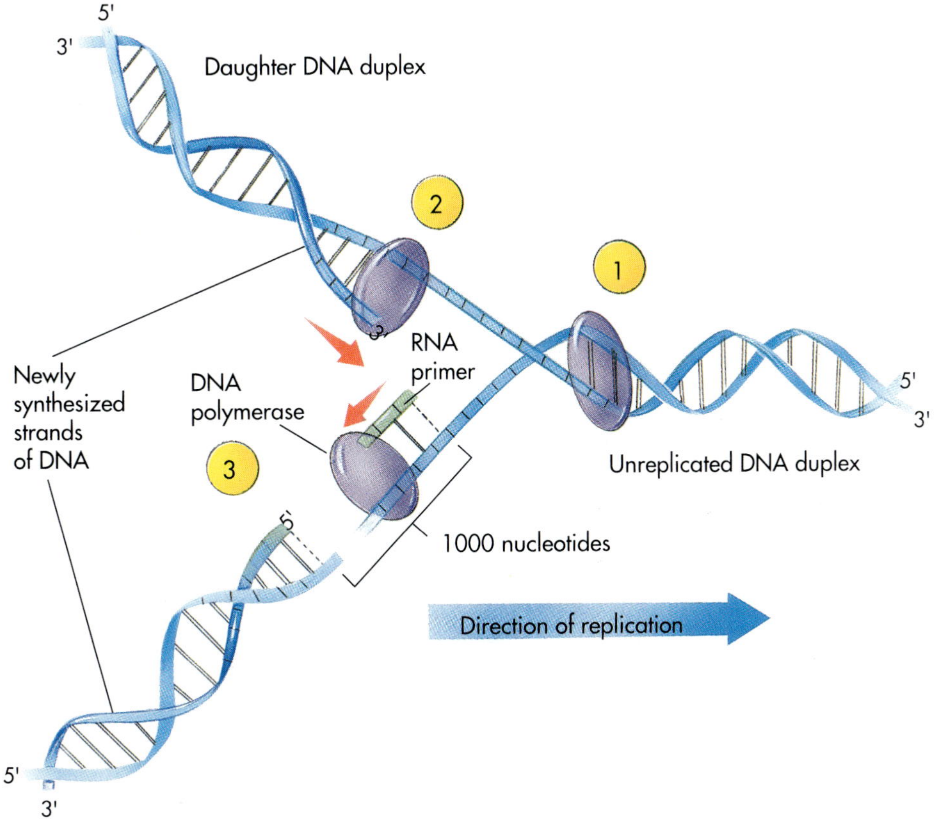

DNA Replication Fork
Figure 14–18

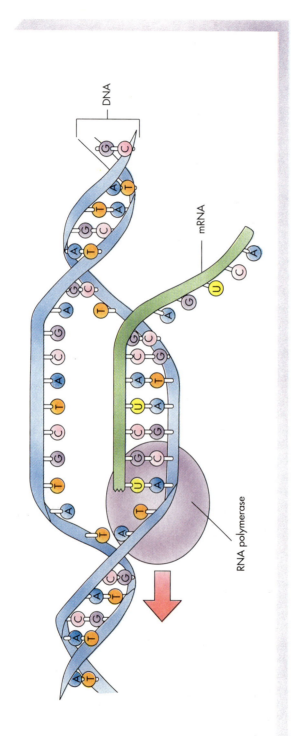

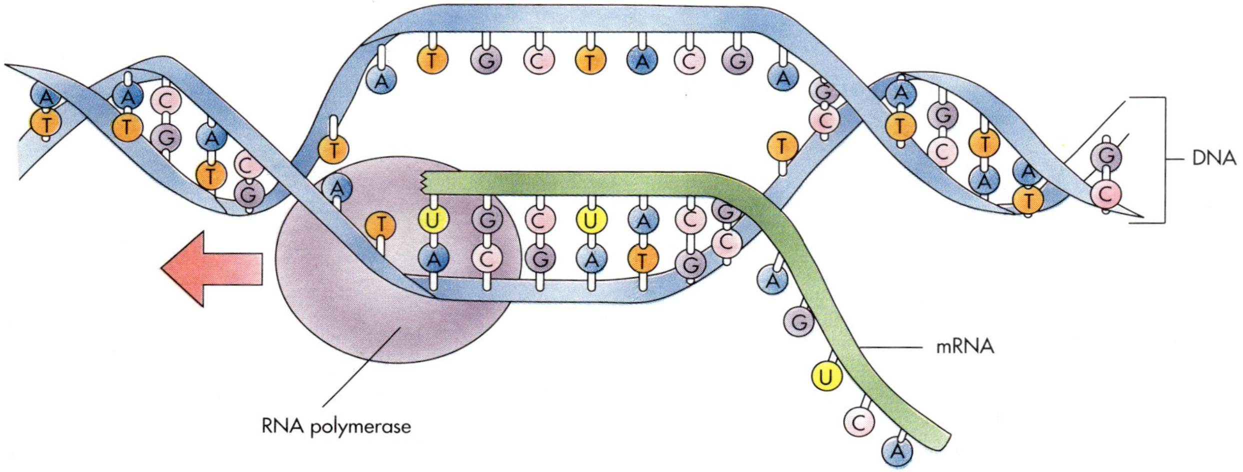

Transcription
Figure 15–4

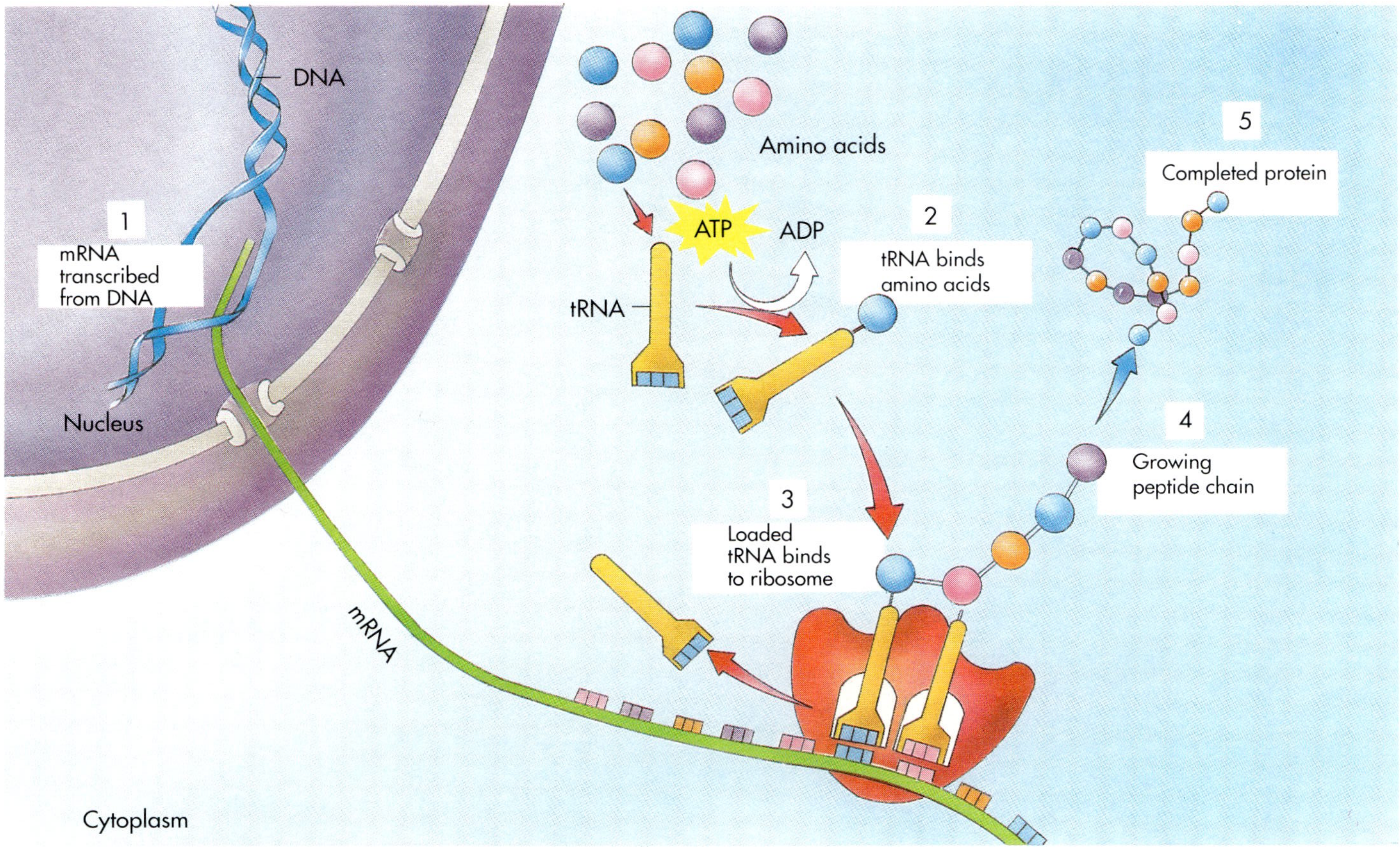

Overview of Protein Synthesis
Figure 15–5

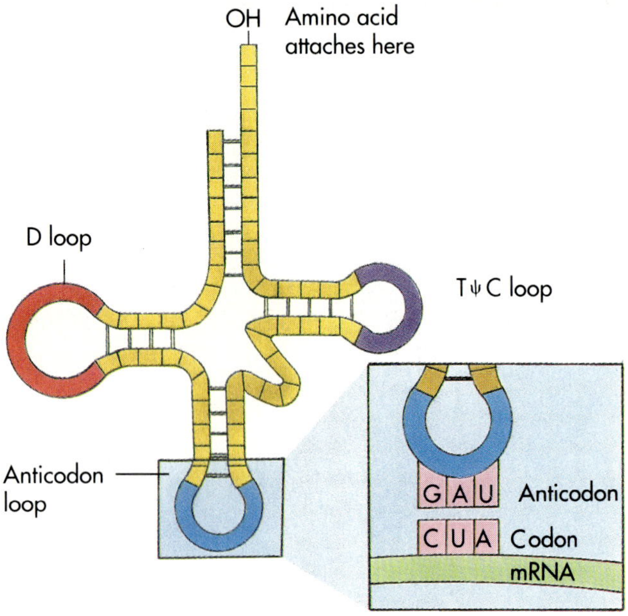

Structure of a tRNA Molecule
Figure 15–8

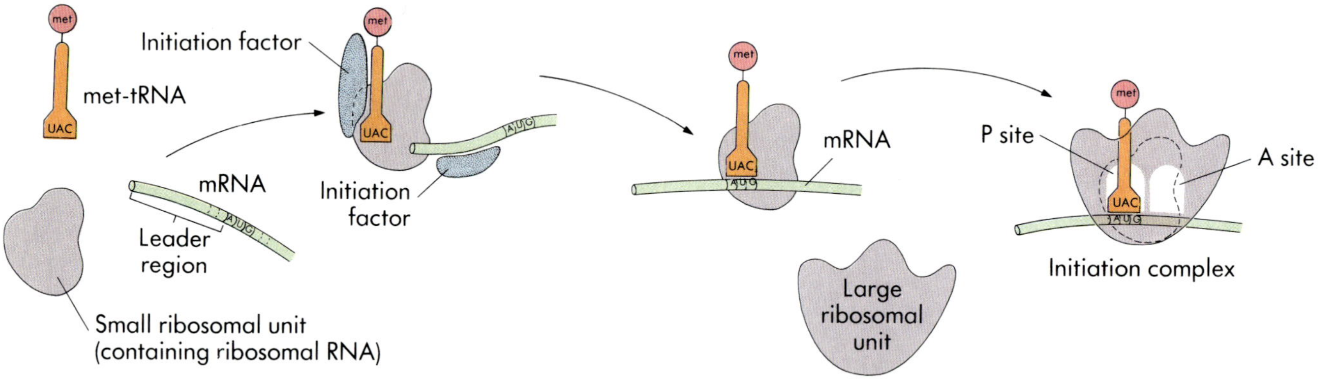

Formation of the Initiation Complex
Figure 15–10

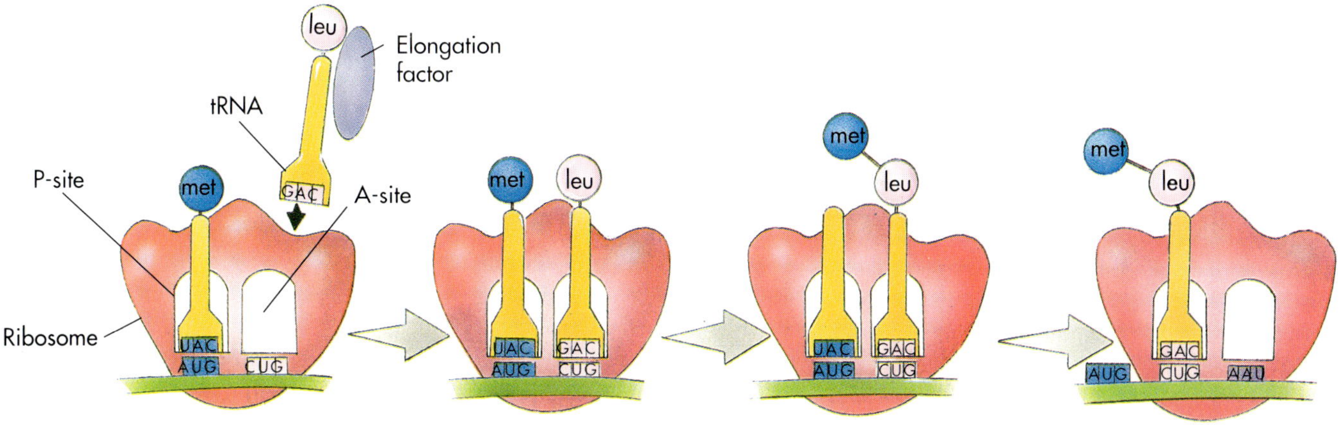

How Polypeptide Synthesis Proceeds
Figure 15–11

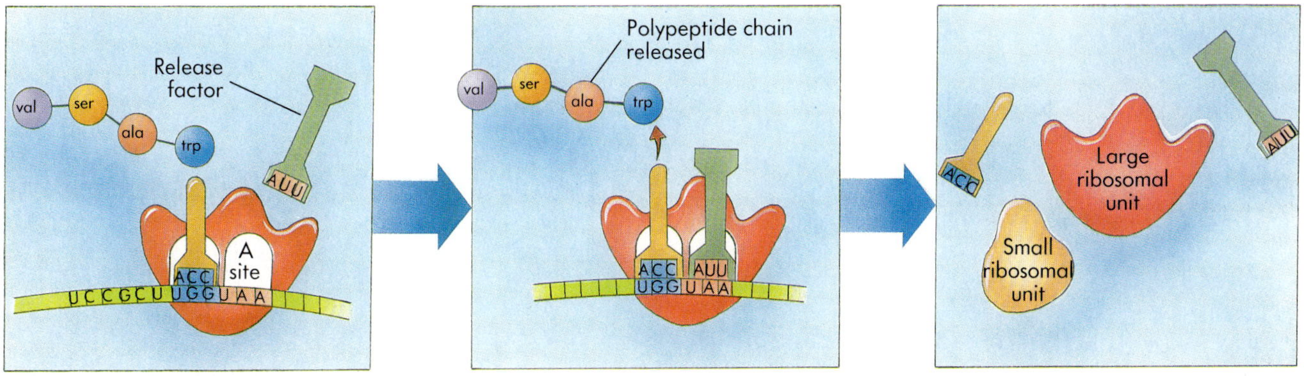

Termination of Protein Synthesis
Figure 15–13

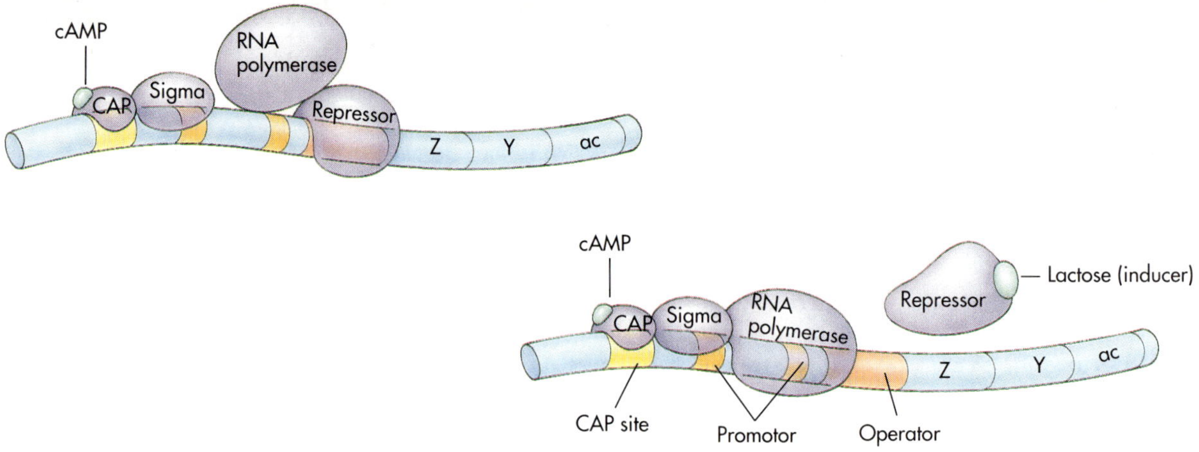

How the *lac* Operon Works
Figure 15–19

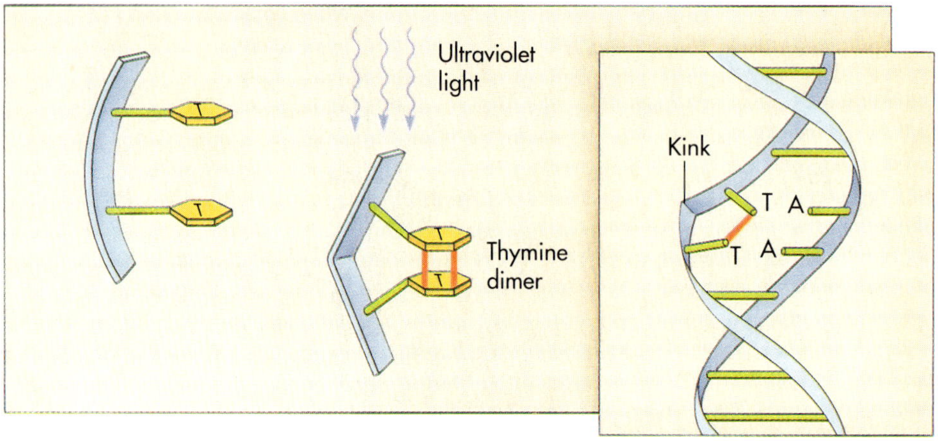

Making a Thymine Dimer
Figure 16–2

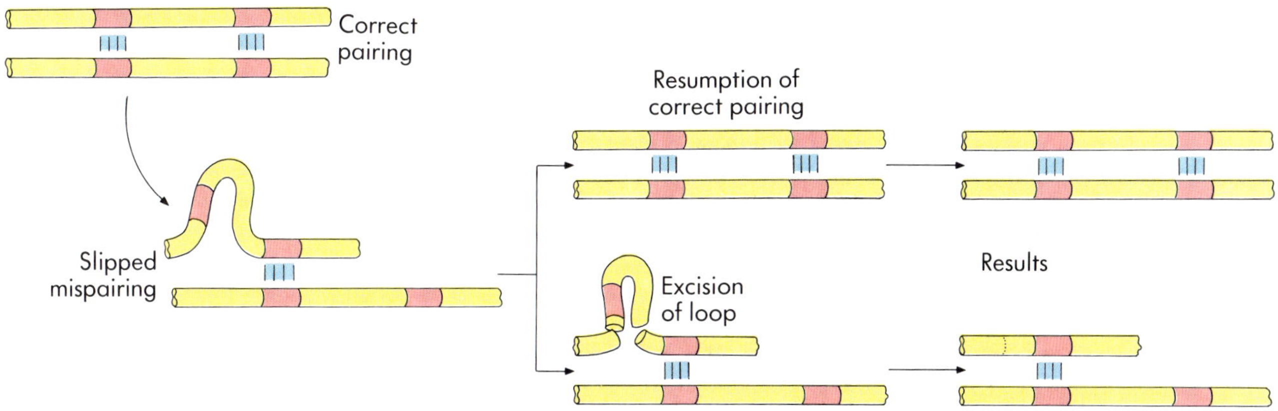

Slipped Mispairing
Figure 16–6

Portrait of a Cancer
Figure 16–8

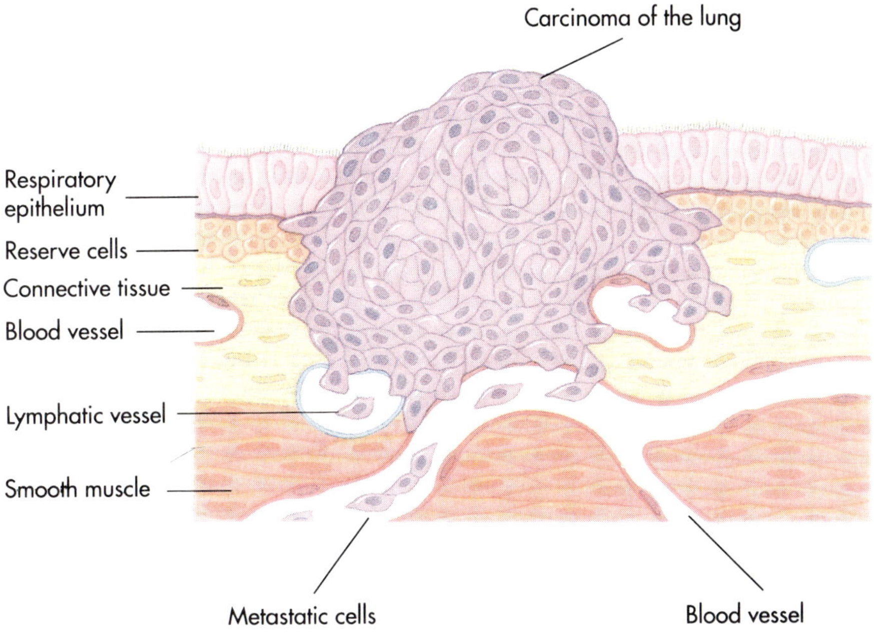

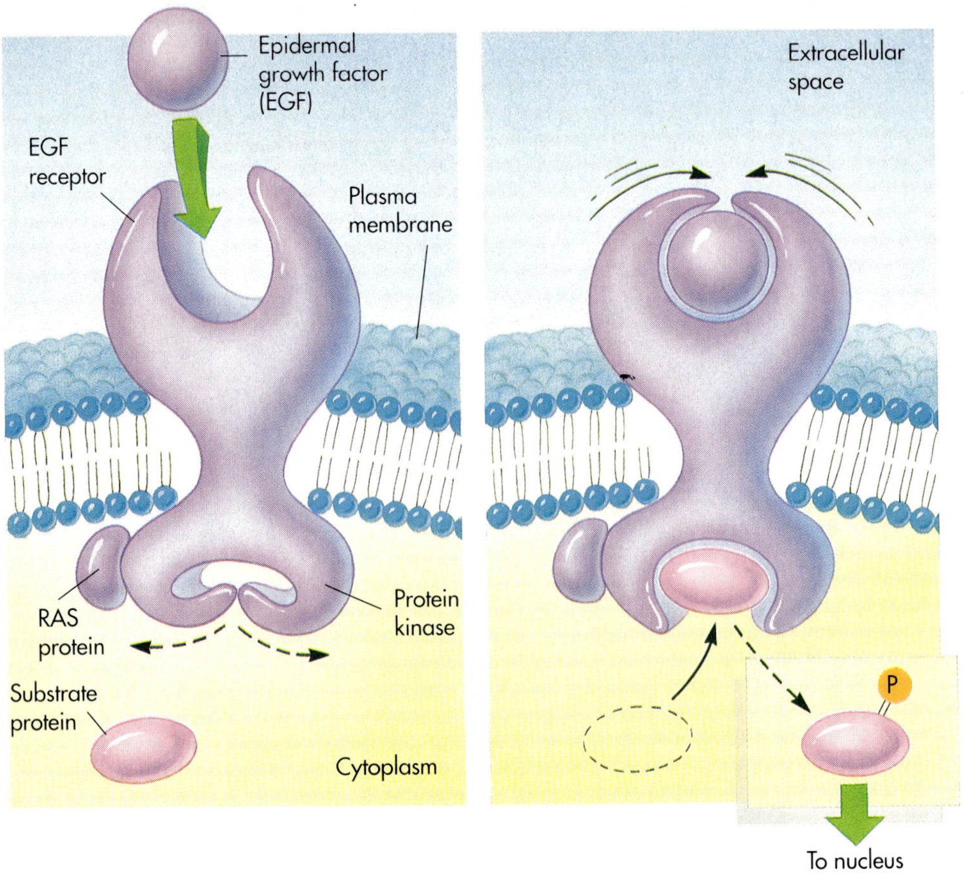

How a Mutation Can Cause Cancer
Figure 16–12

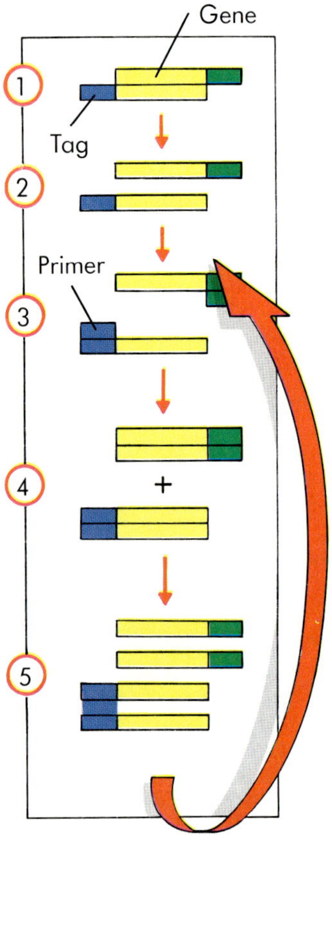

Polymerase Chain Reaction
Figure 18–8

Phenotypes			
Genotypes	BB	Bb	bb
Frequency of genotype in population	0.36	0.48	0.16
Frequency of gametes	0.36 + 0.24 = 0.6B		0.24 + 0.16 = 0.4b

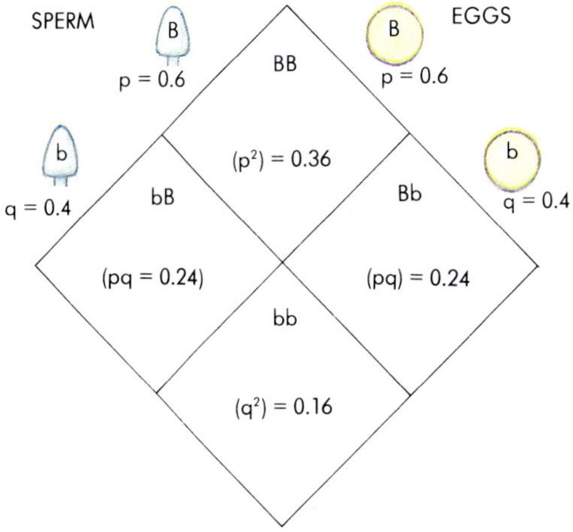

Hardy-Weinberg Equilibrium
Figure 19–3

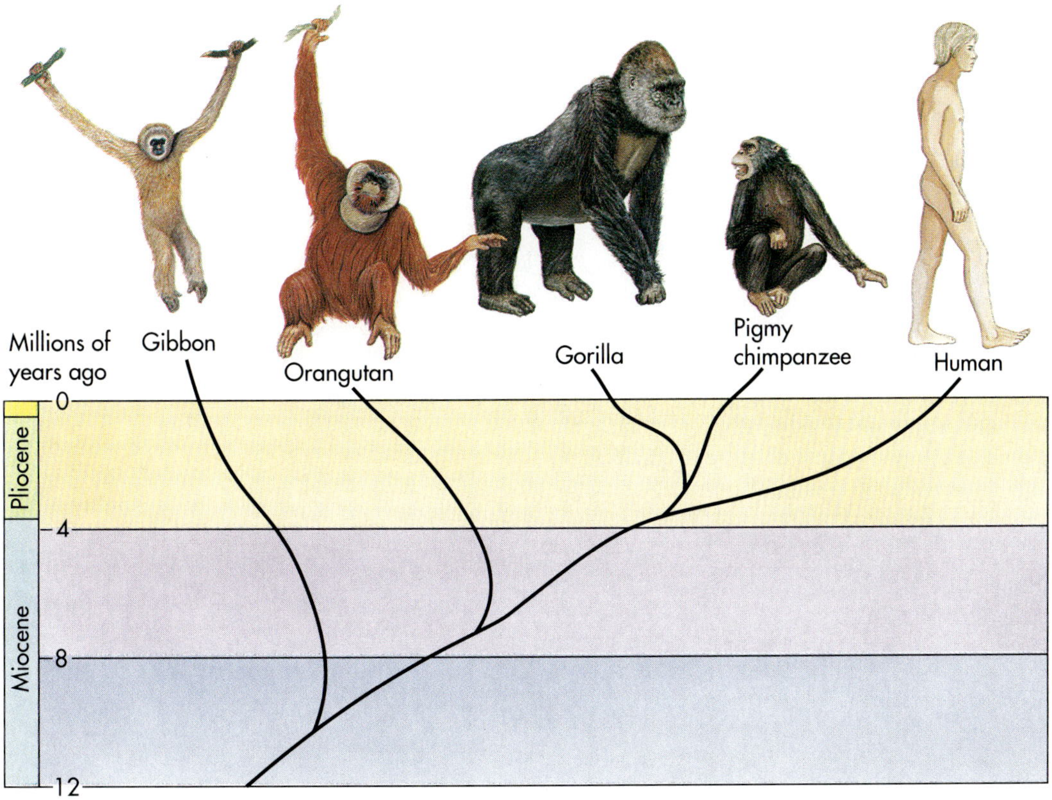

Evolution of Living Hominoids
Figure 22–22

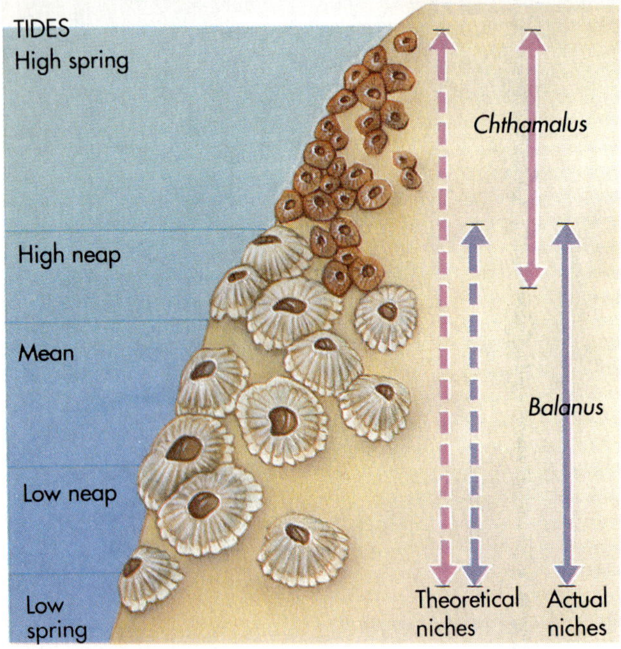

Competition Among Barnacles
Figure 23–12

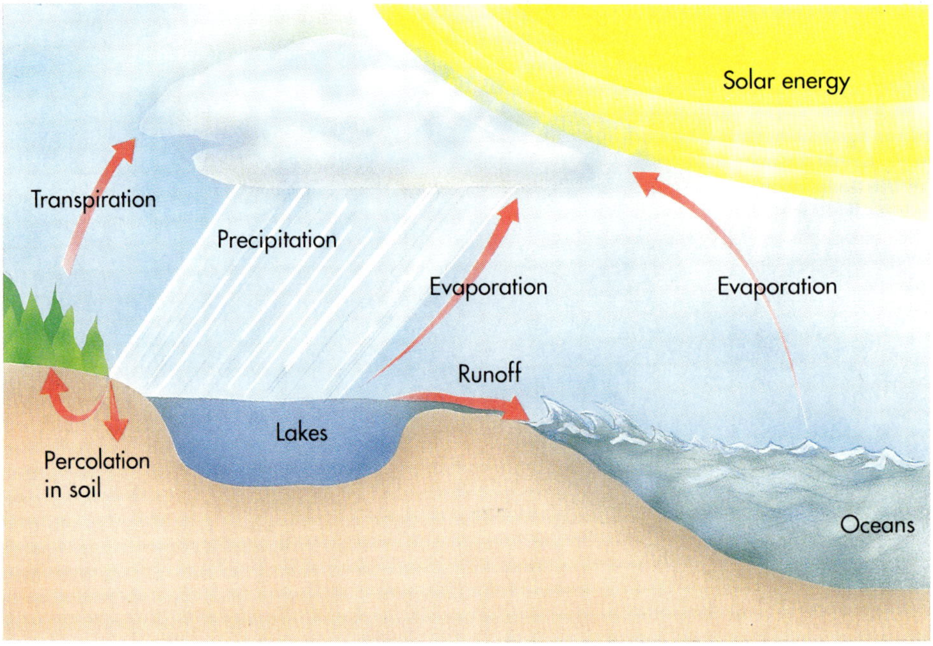

Water Cycle
Figure 25–2

Carbon Cycle
Figure 25–3

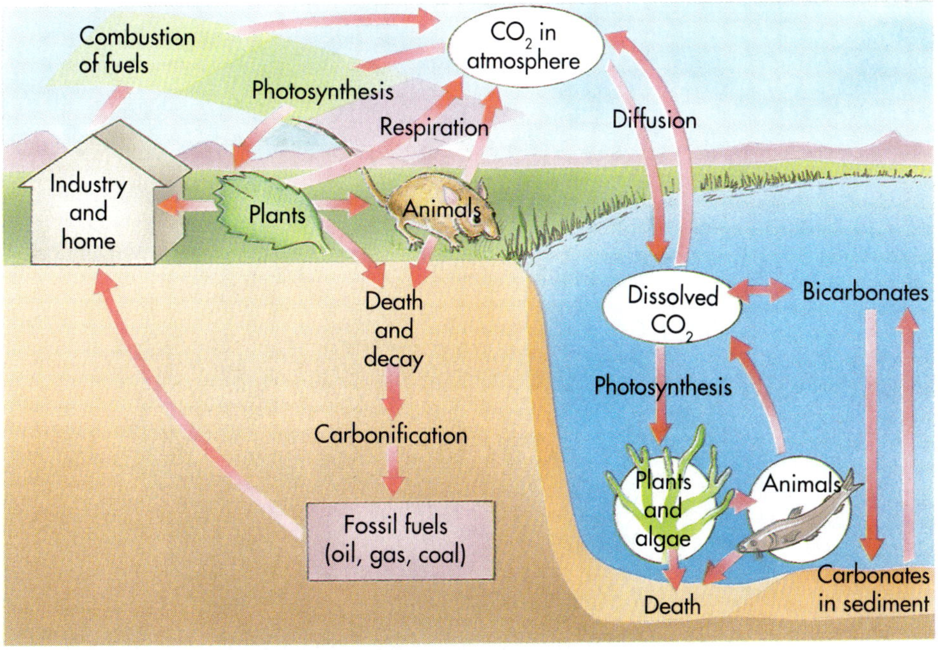

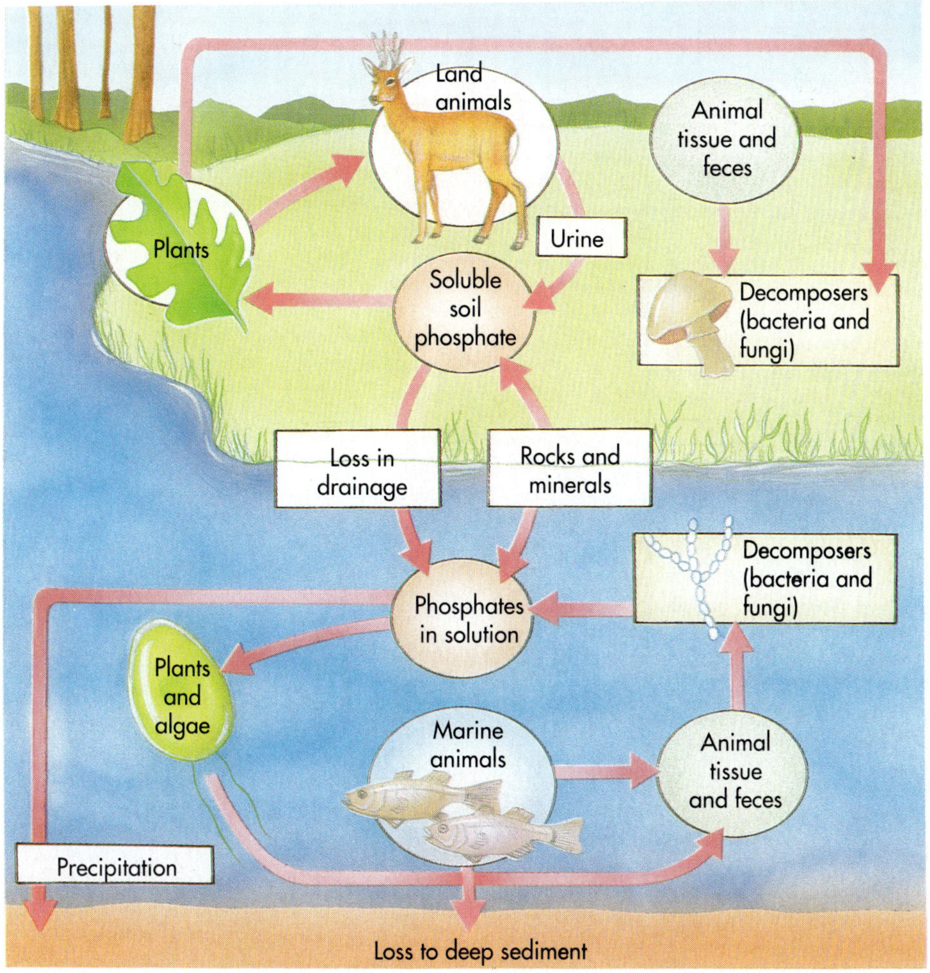

Phosphorus Cycle
Figure 25–8

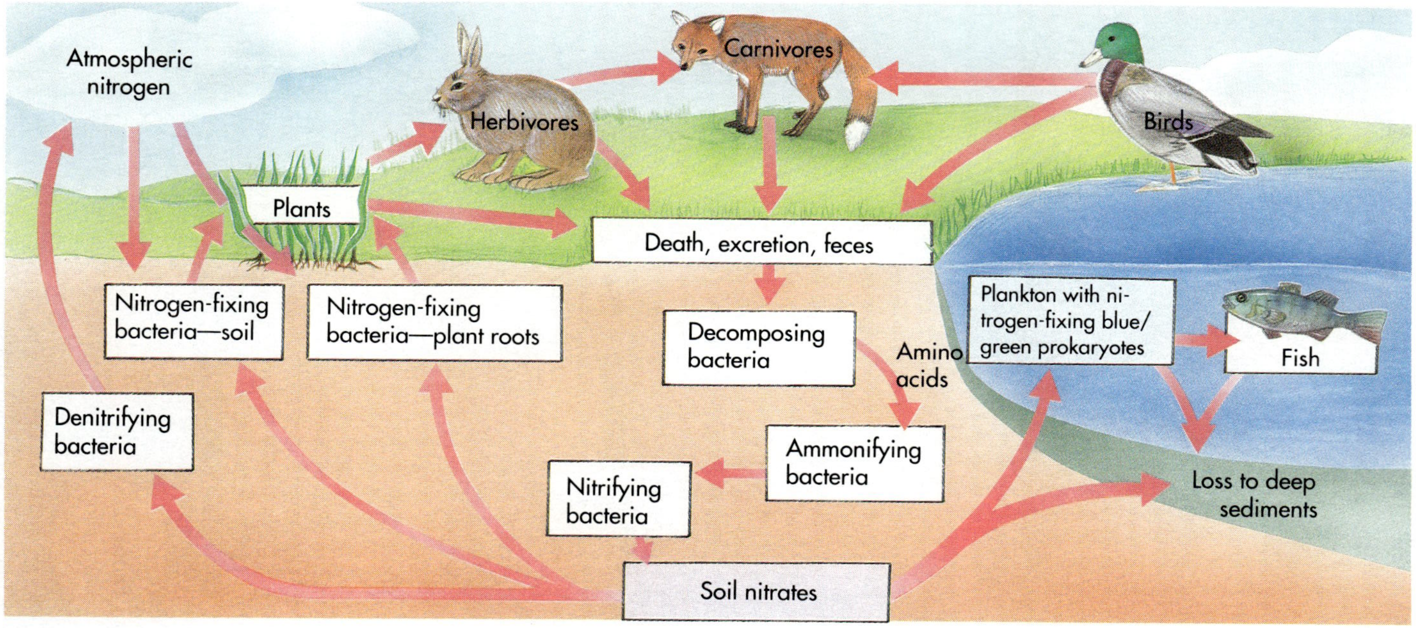

Nitrogen Cycle
Figure 25–4

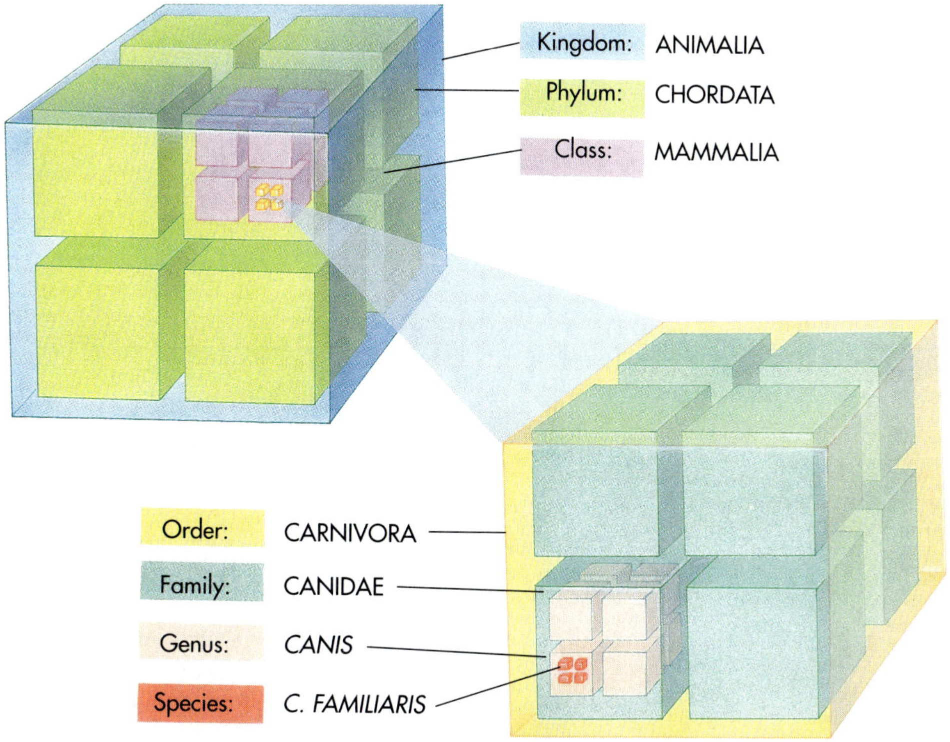

Hierarchical System Used in Classifying Organisms
Figure 28–4

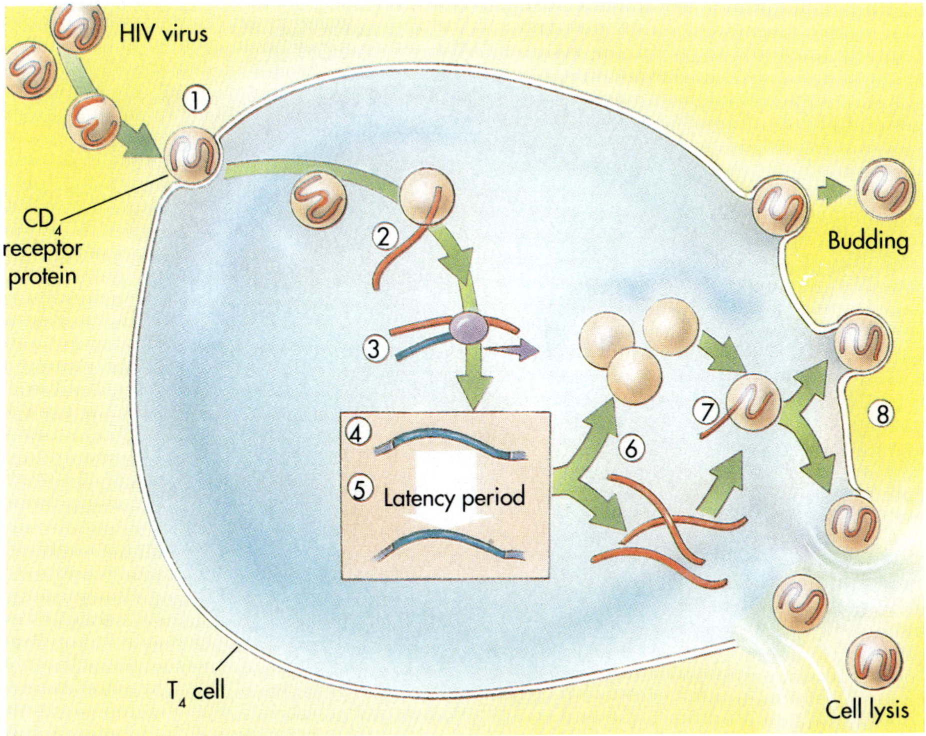

HIV Infection Cycle
Figure 29–8

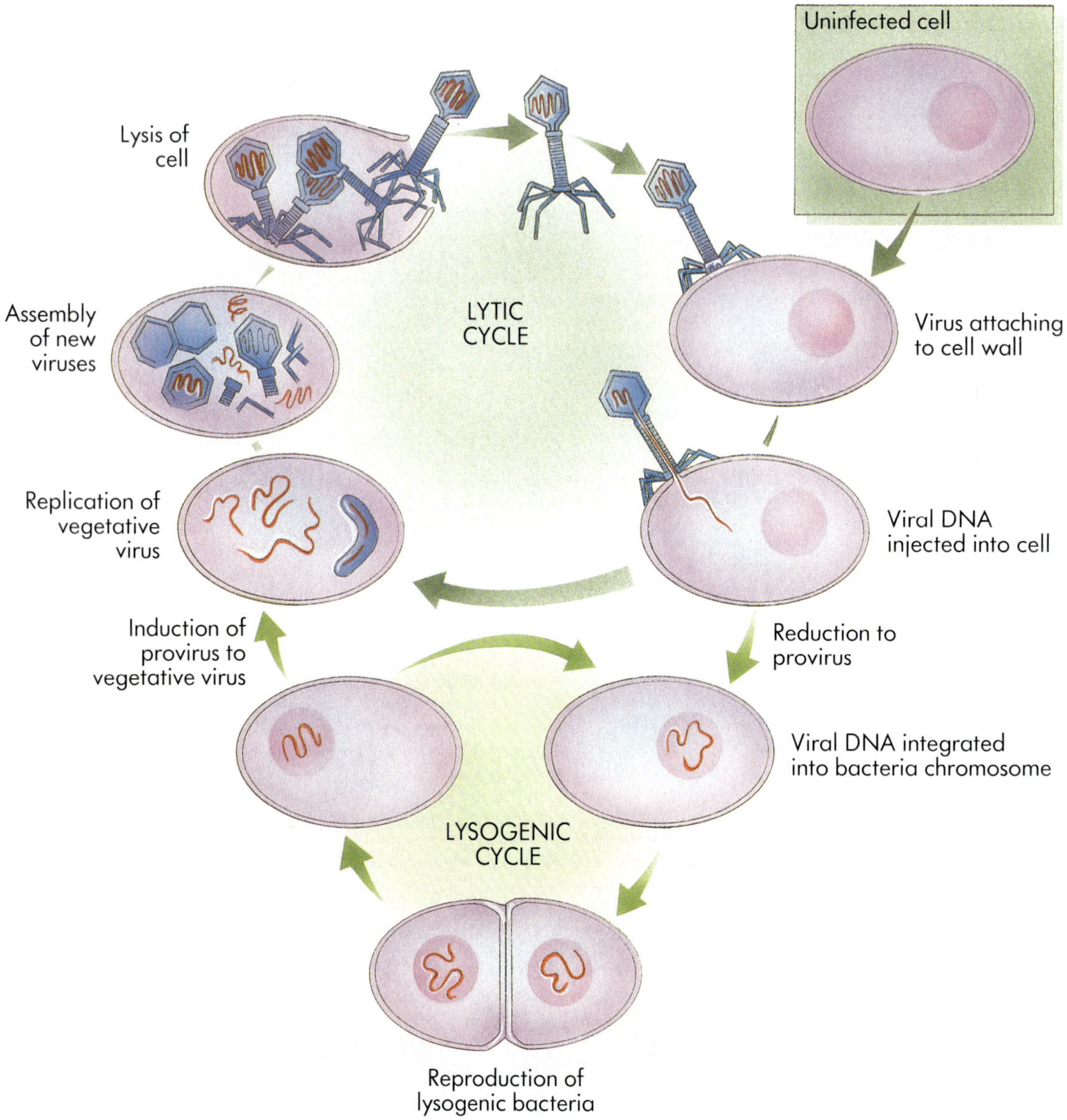

Lytic and Lysogenic Cycles of a Bacteriophage
Figure 29–10

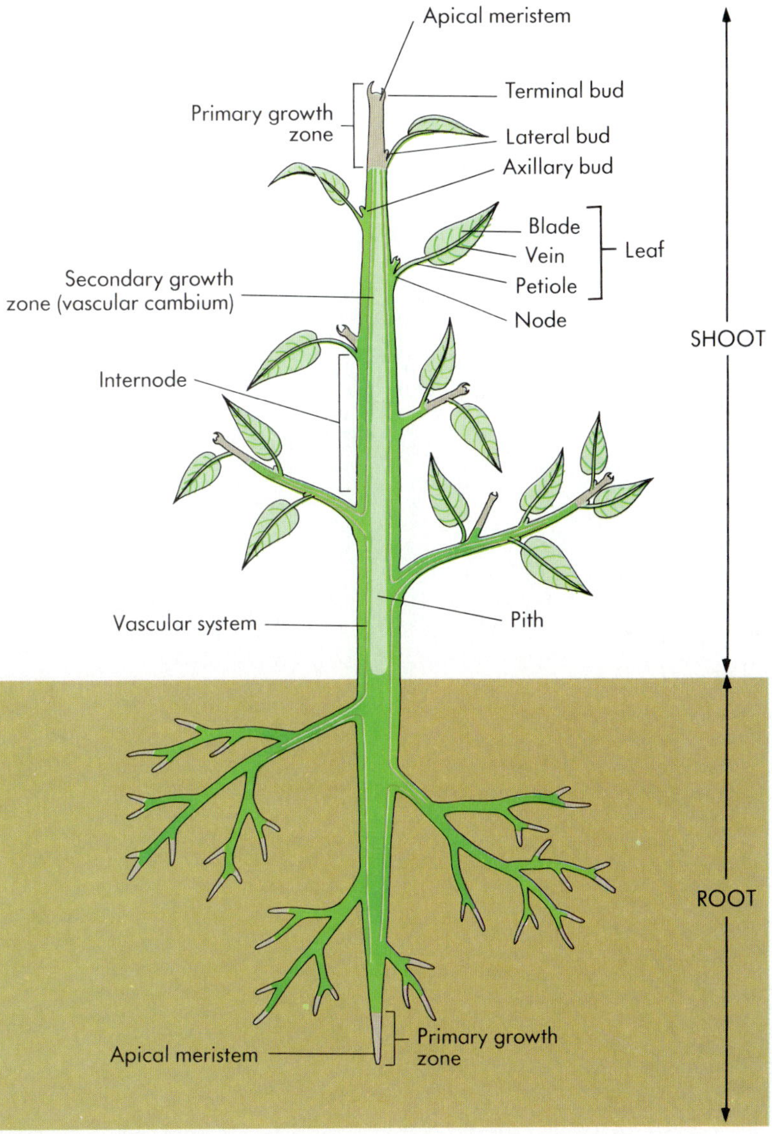

Dicot Plant Body
Figure 35–1

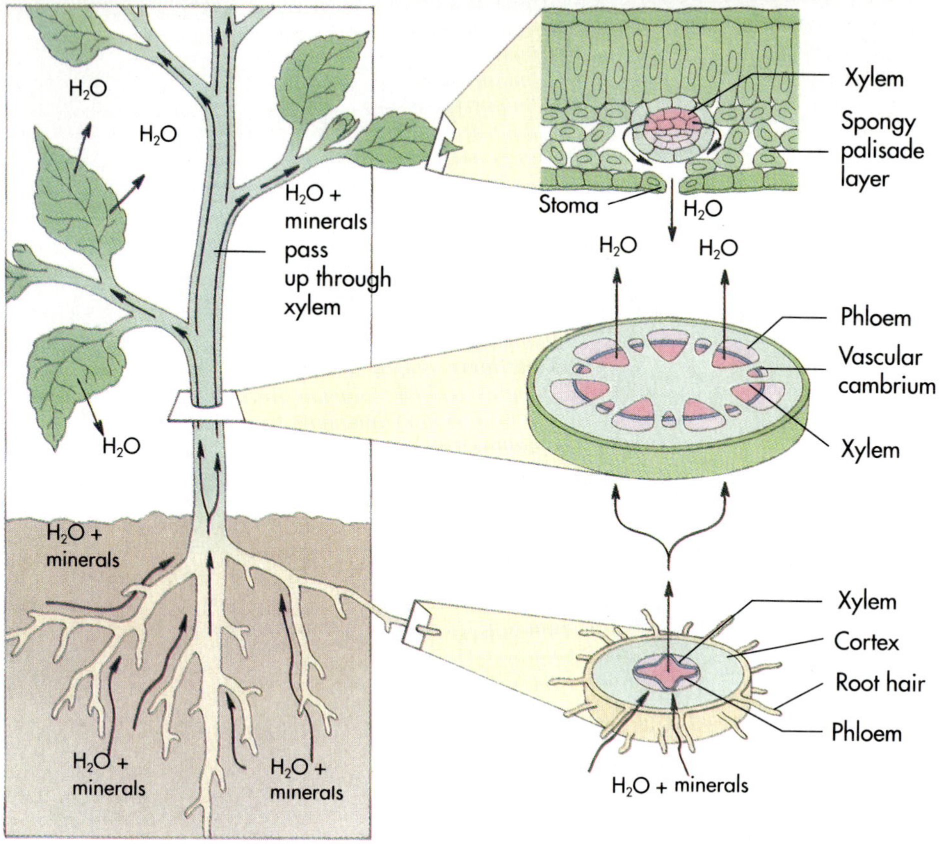

Flow of Plant Materials
Figure 36–1

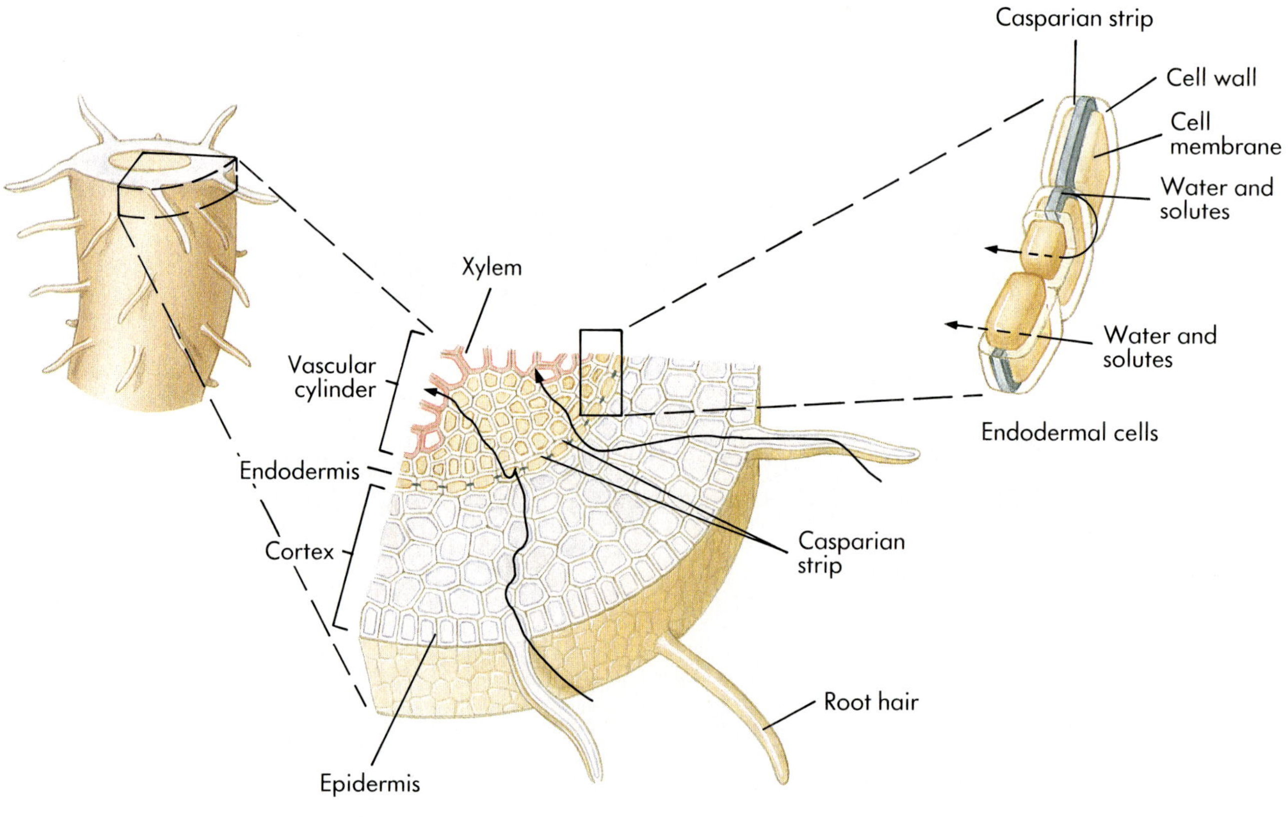

Pathways of Mineral Transport in Roots
Figure 36–4

Basic Sponge Structure
Figure 39–4

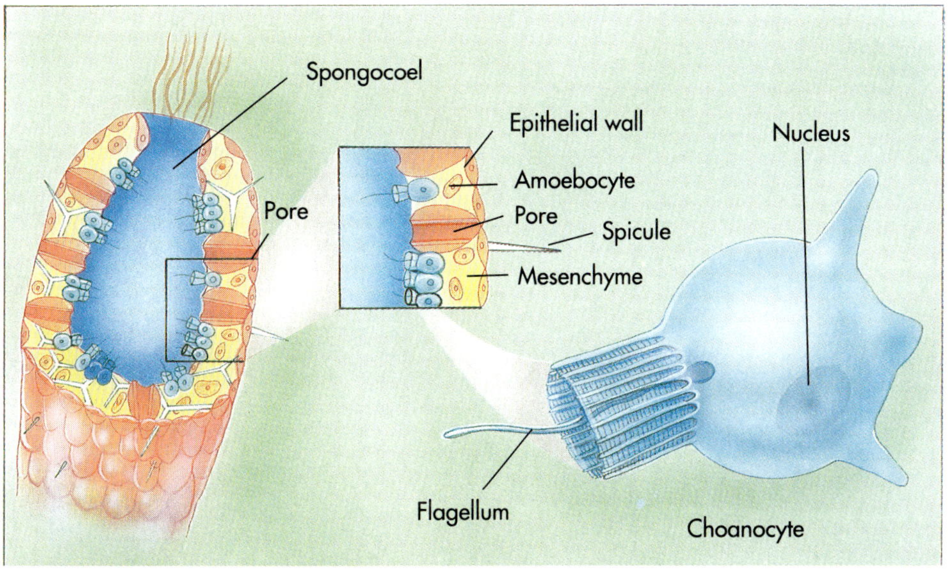

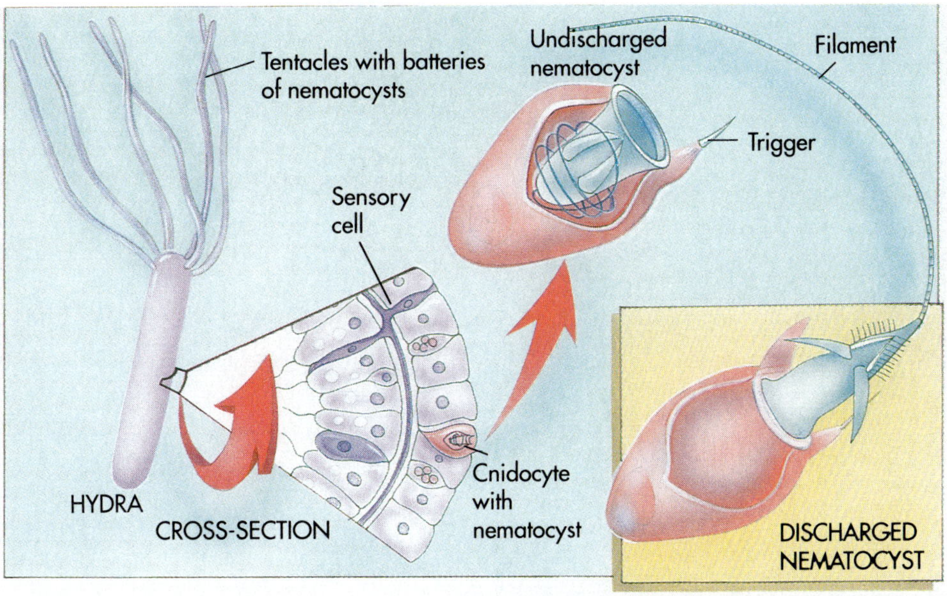

Basic *Hydra* Structure
Figure 39–9

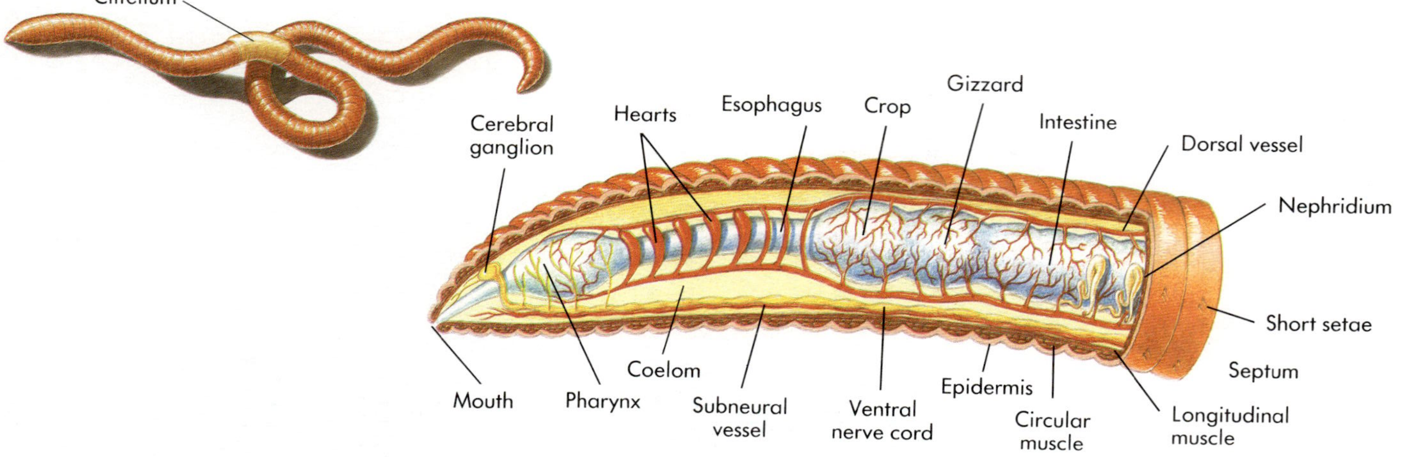

Earthworm
Figure 40–16

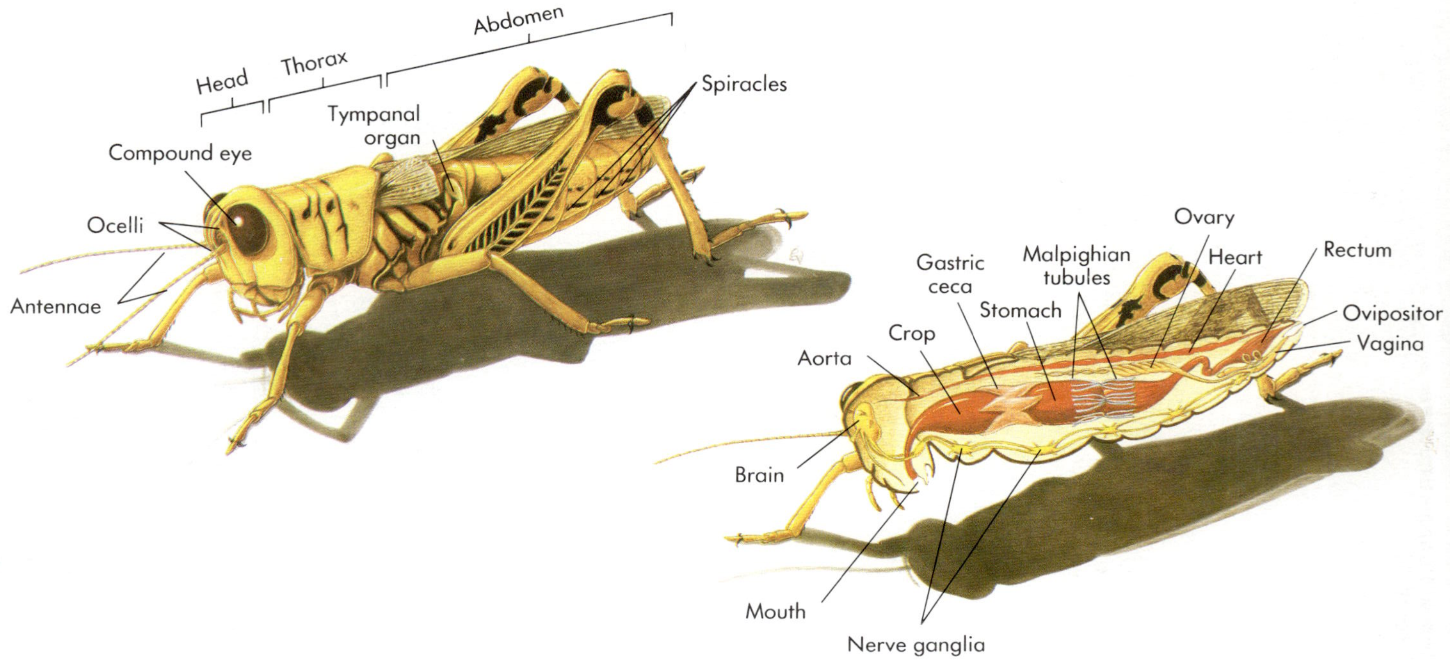

Grasshopper
Figure 41–7

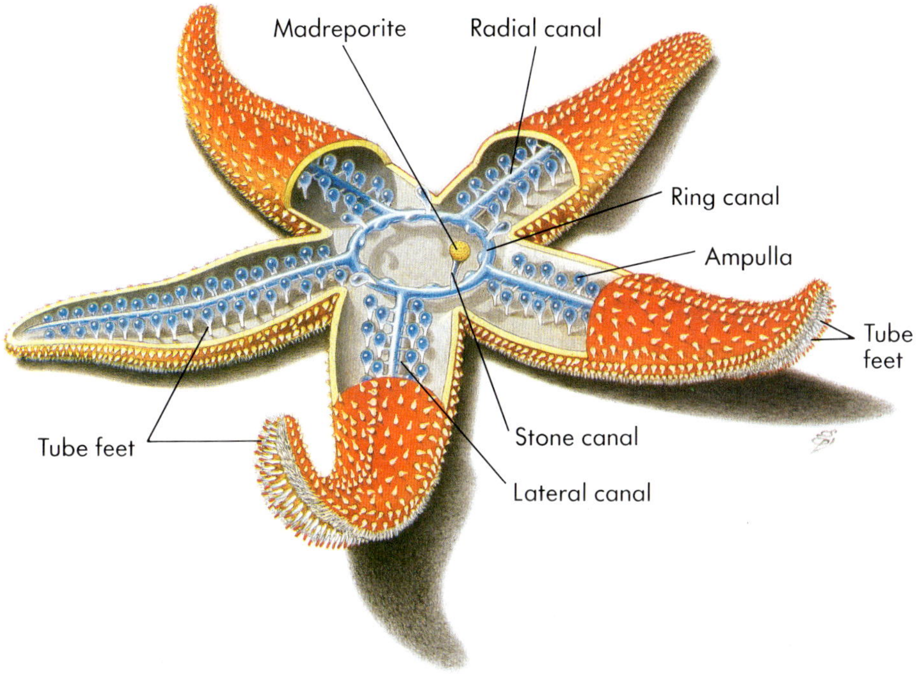

Echinoderm Structure
Figure 42–2

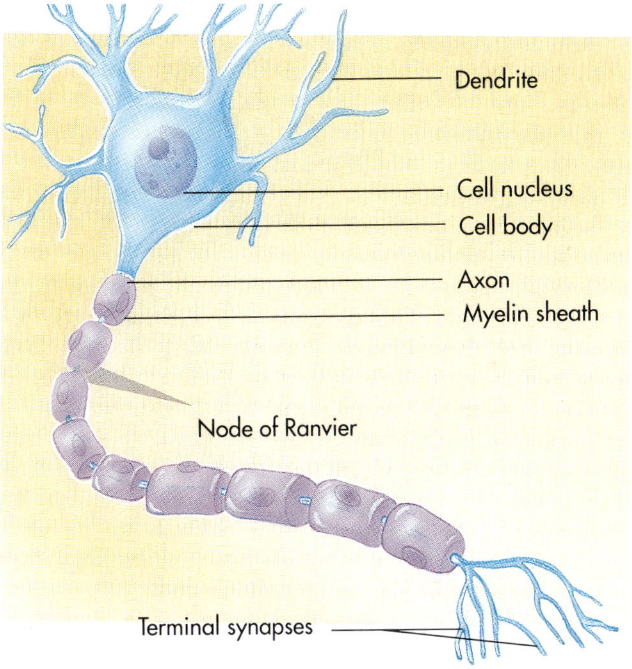

Idealized Structure of a Vertebrate Neuron
Figure 43–13

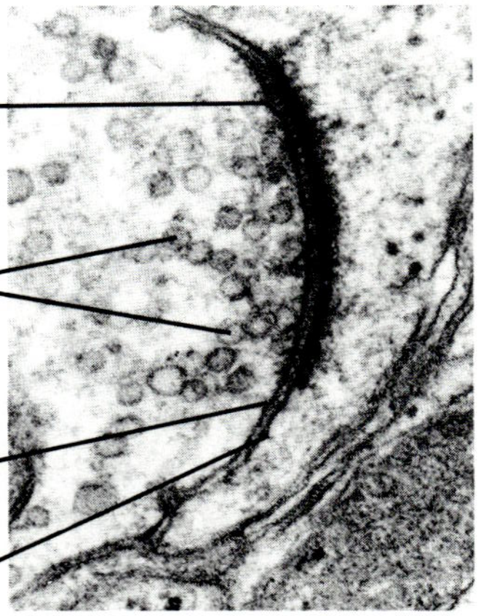

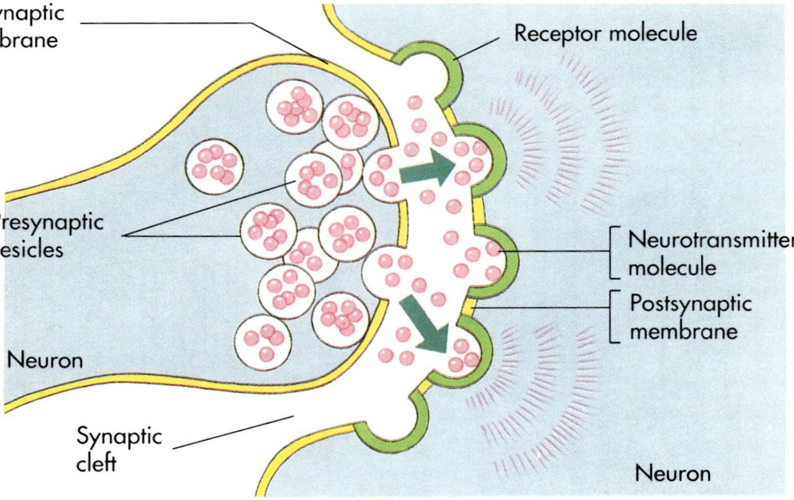

Synaptic Cleft Between Two Neurons
Figure 44–13

Photo by Dr. Lennart Heimer.

Section Through the Human Brain
Figure 45–8

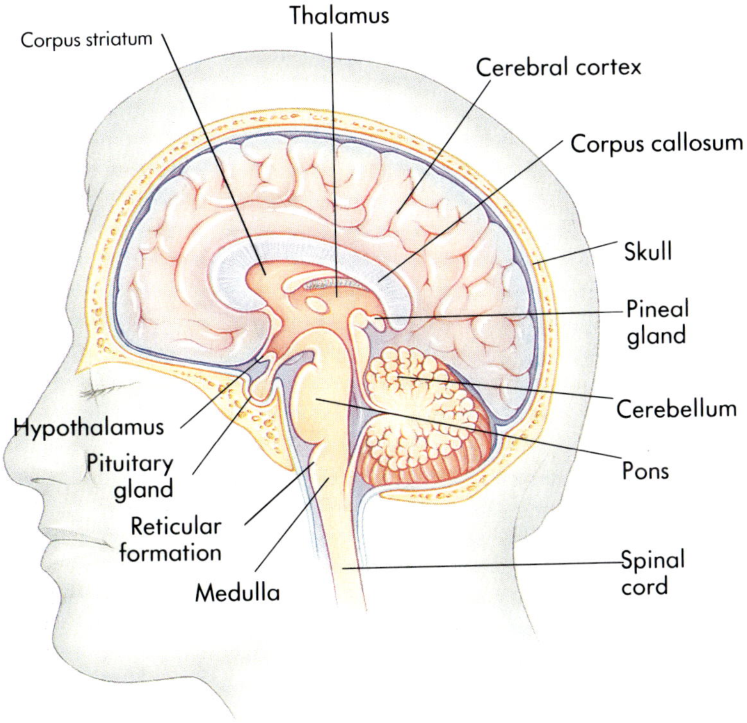

Human Ear Structure
Figure 46–14

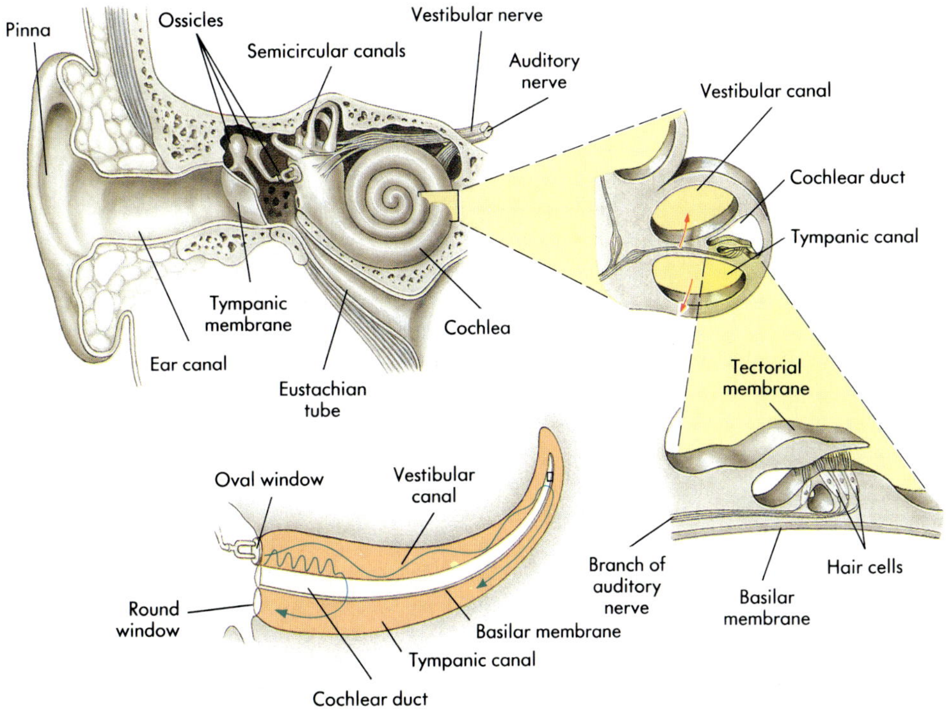

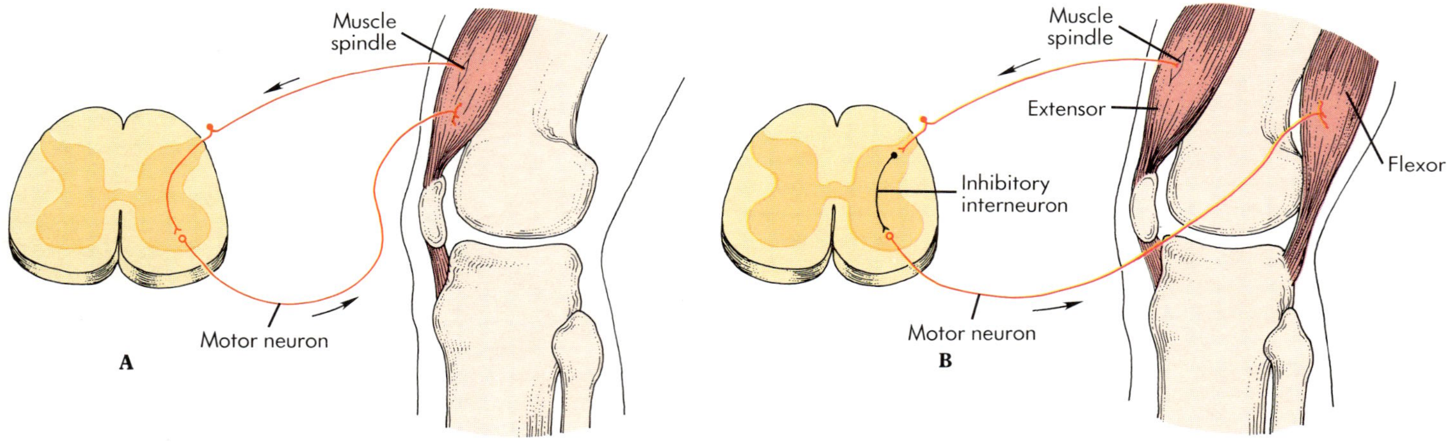

Monosynaptic Reflexes
Figure 45–16

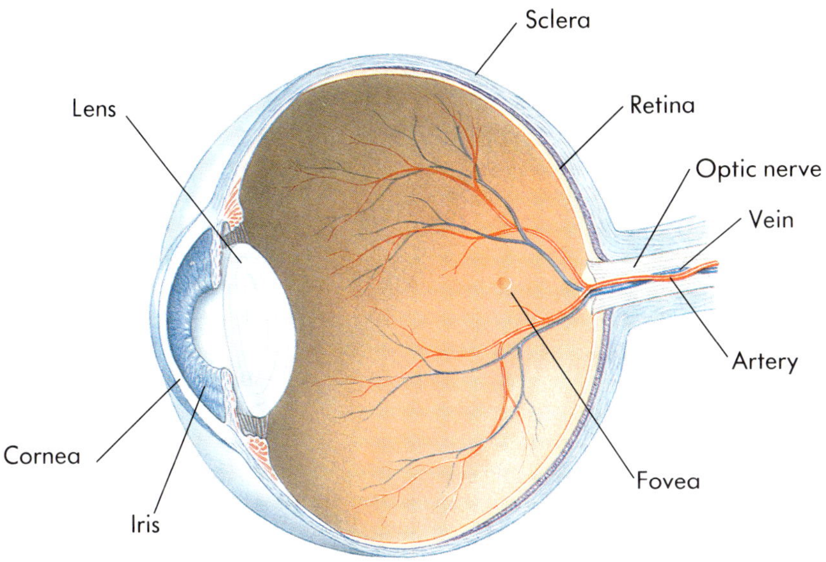

Human Eye Structure
Figure 46–17

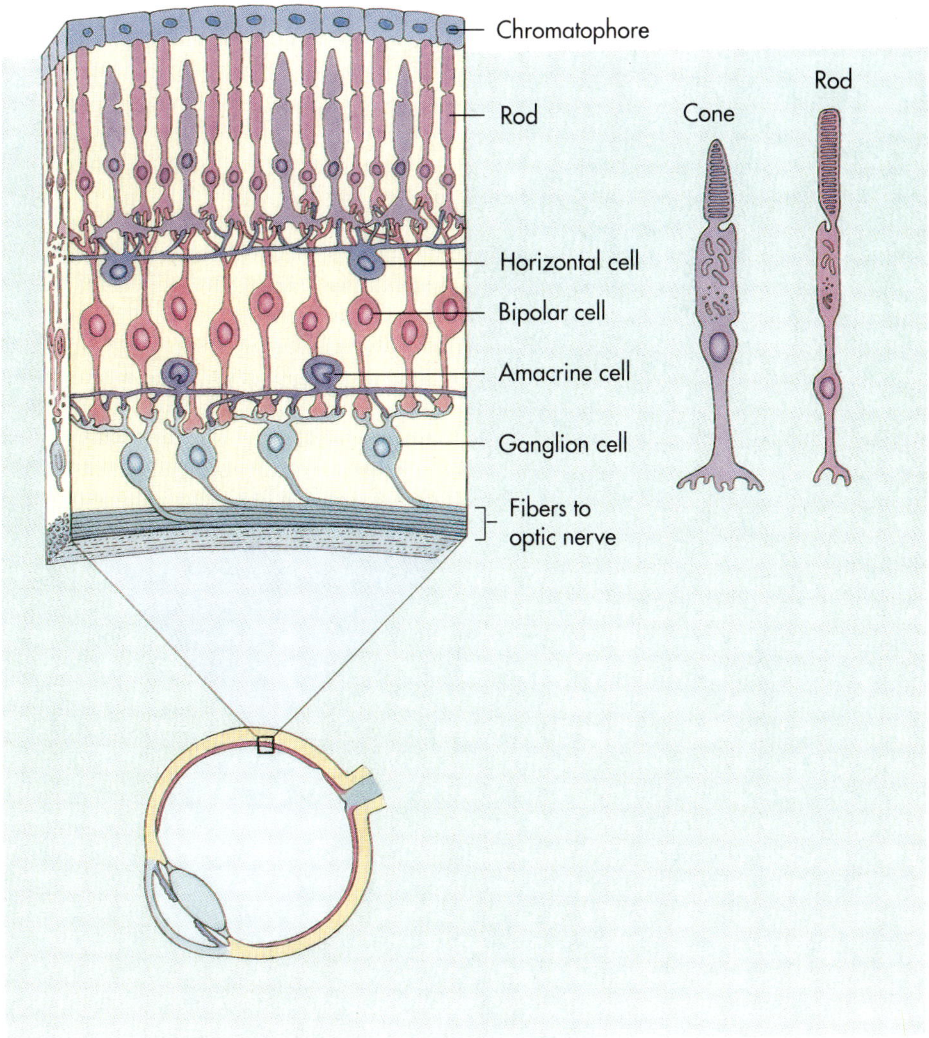

Retina Structure
Figure 46–19

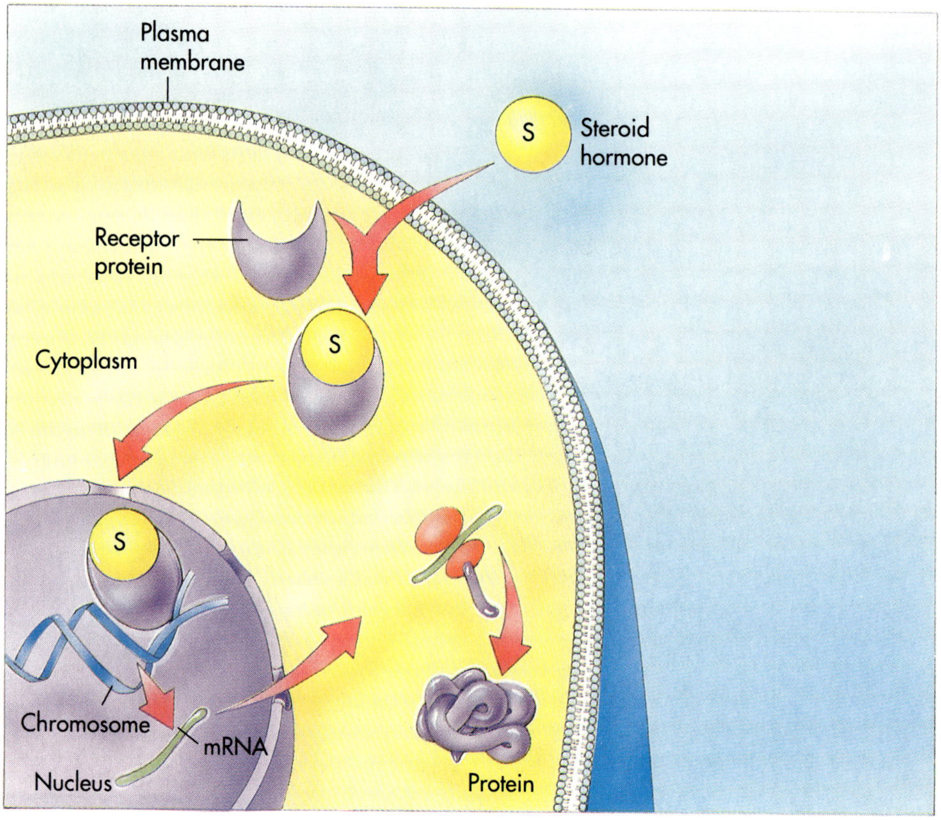

How Steroid Hormones Work
Figure 47–6

How Peptide Hormones Work
Figure 47–7

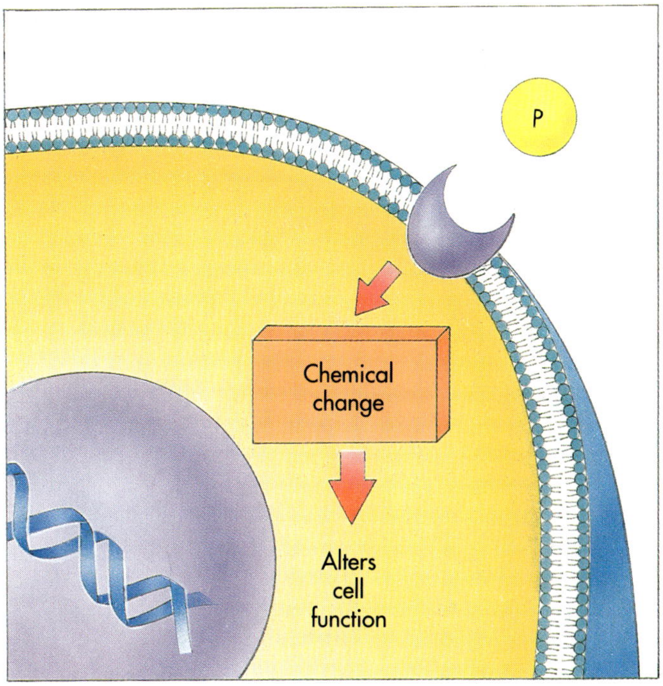

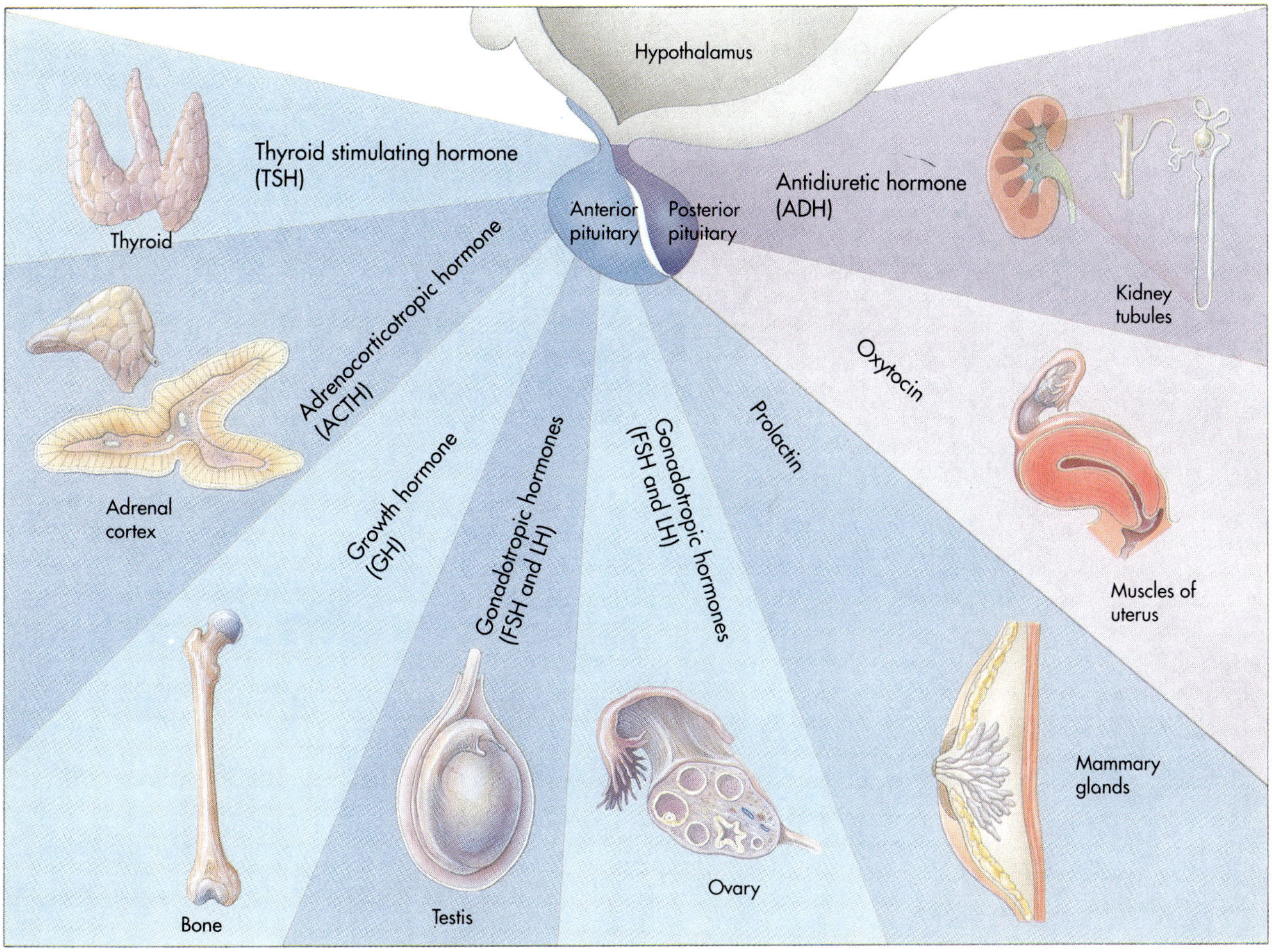

Role of the Pituitary
Figure 47–11

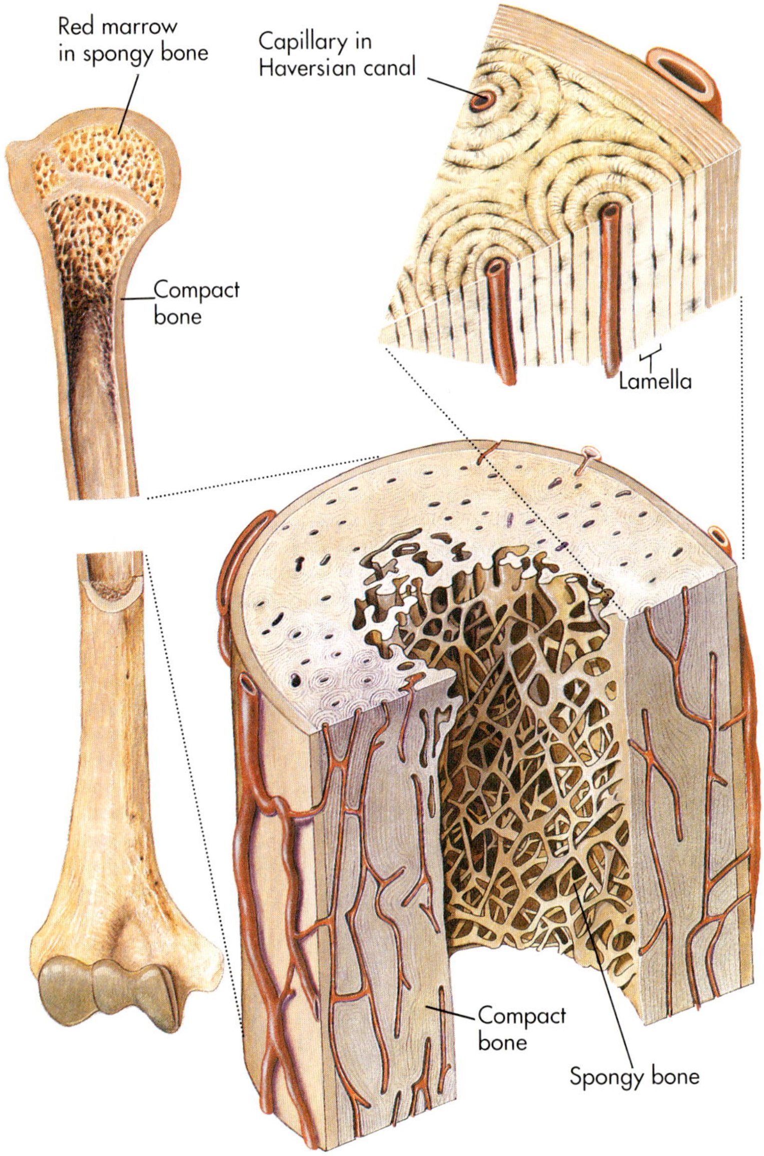

Organization of Compact Bone
Figure 48–2

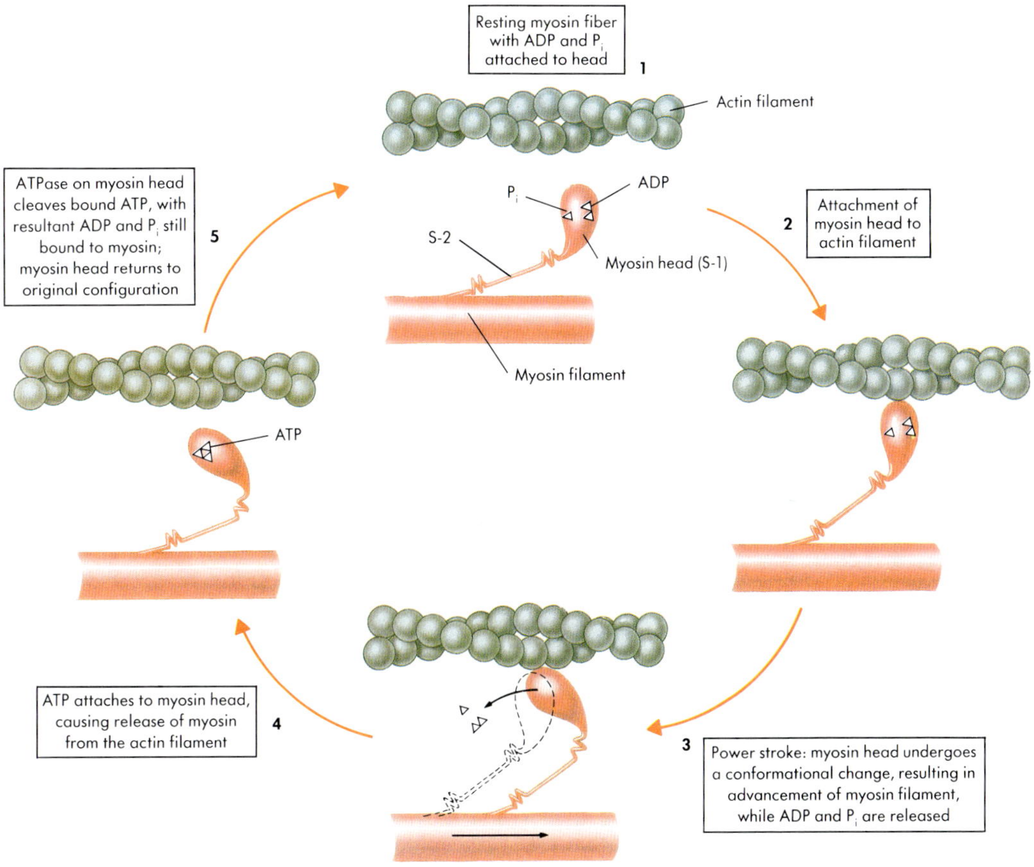

Resting myosin fiber with ADP and P_i attached to head
1
Actin filament
ADP
P_i
S-2
Myosin head (S-1)
Myosin filament
2
Attachment of myosin head to actin filament
3
Power stroke: myosin head undergoes a conformational change, resulting in advancement of myosin filament, while ADP and P_i are released
4
ATP attaches to myosin head, causing release of myosin from the actin filament
ATP
5
ATPase on myosin head cleaves bound ATP, with resultant ADP and P_i still bound to myosin; myosin head returns to original configuration

Mechanisms of Myofilament Contraction
Figure 48–13

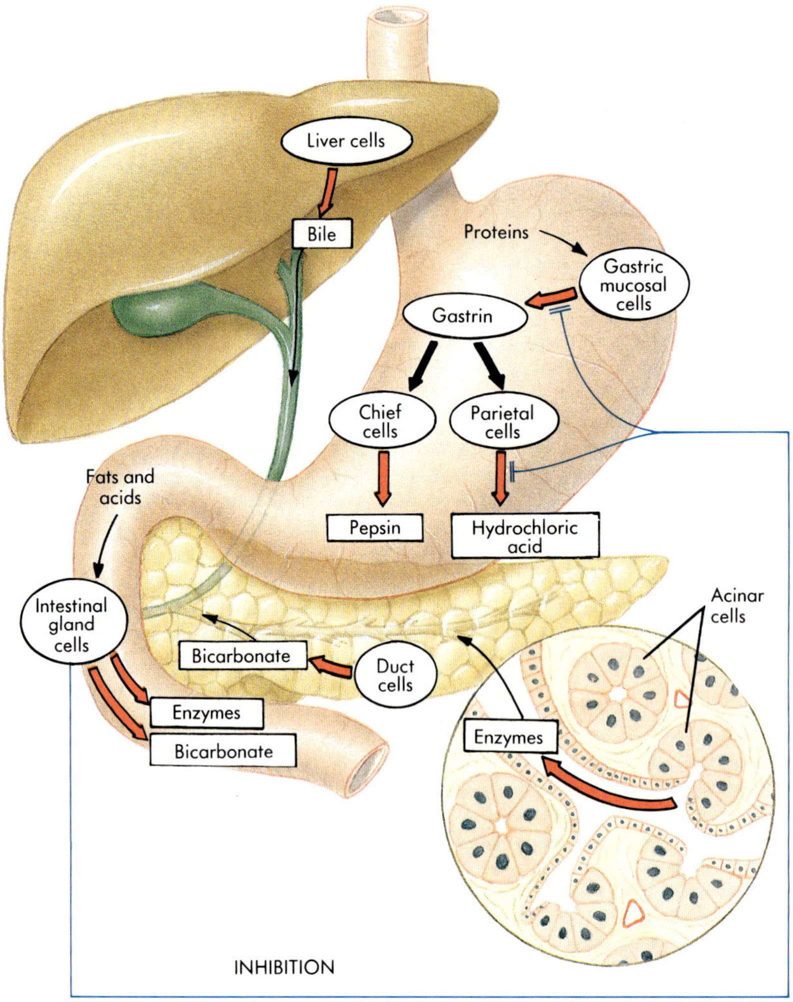

Regulation of Digestive Enzyme Production
Figure 49–11

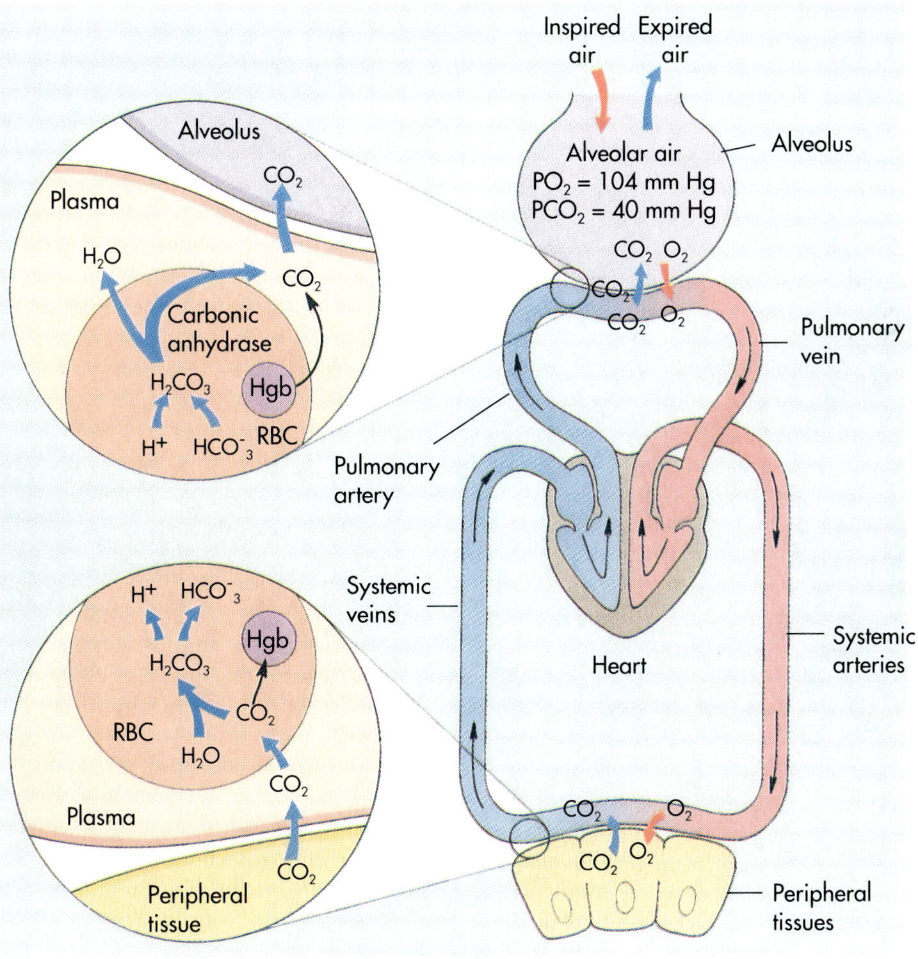

Respiratory Journey
Figure 50–19

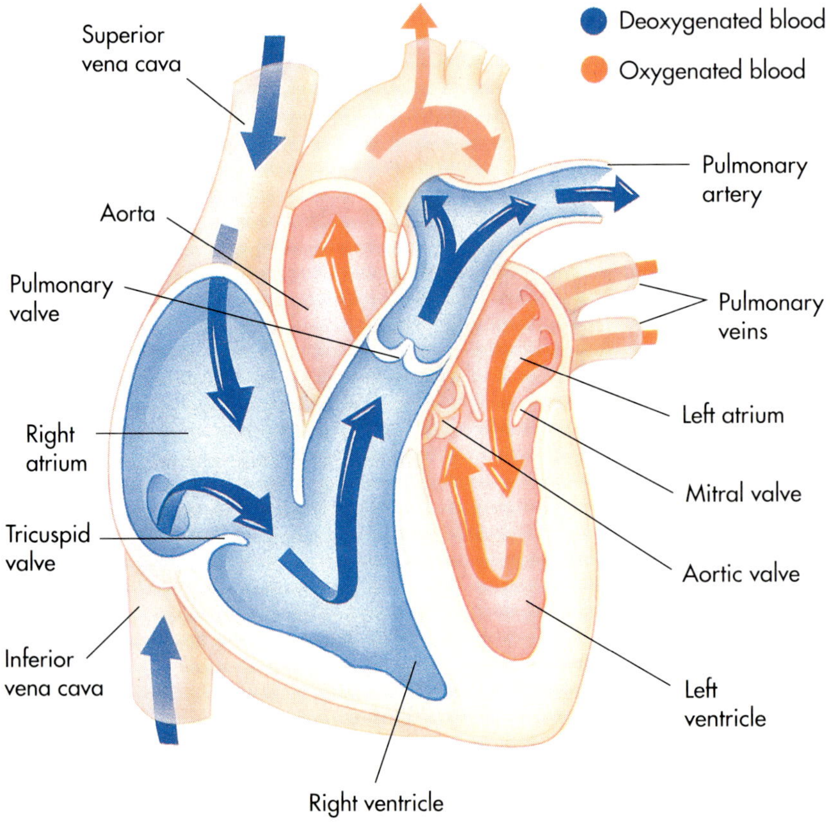

Path of Blood Through the Heart
Figure 51–20

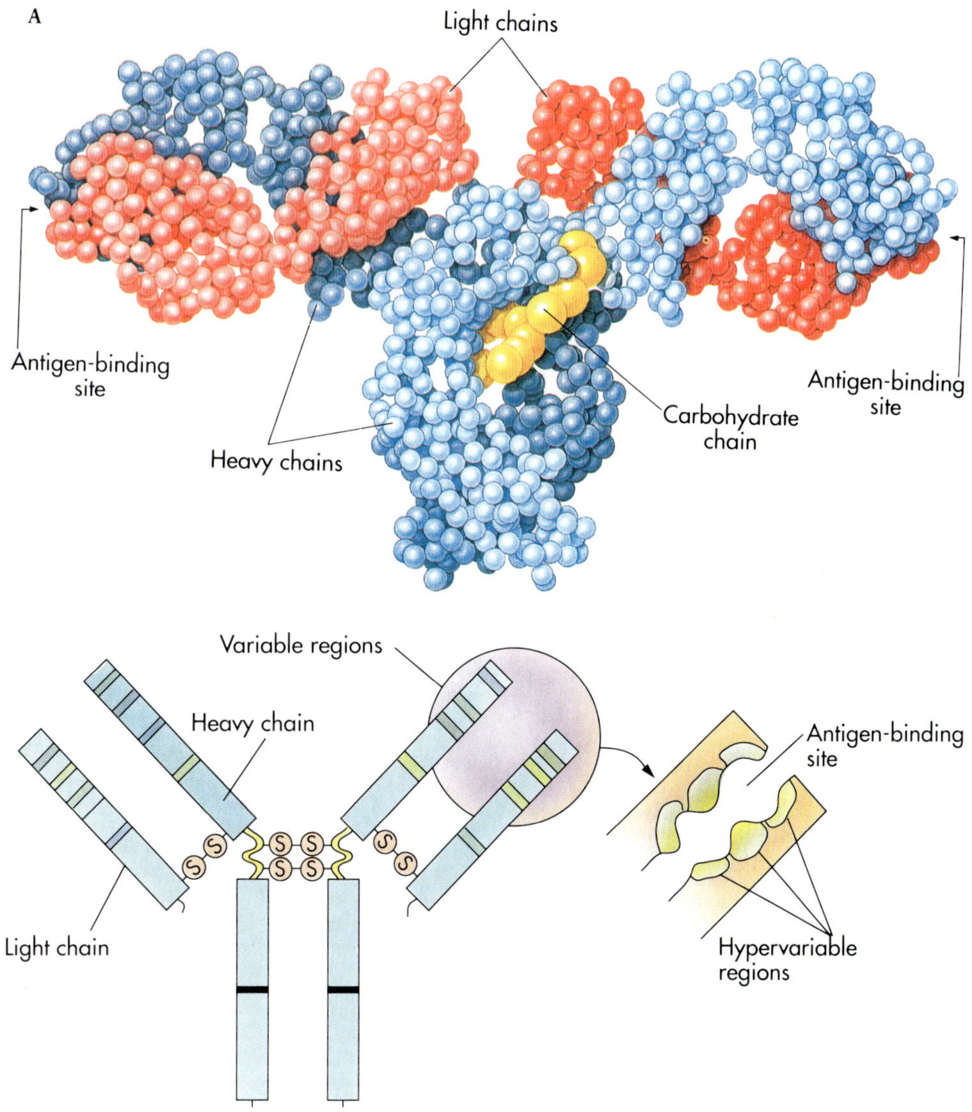

Structure of an Antibody Molecule
Figure 52–25

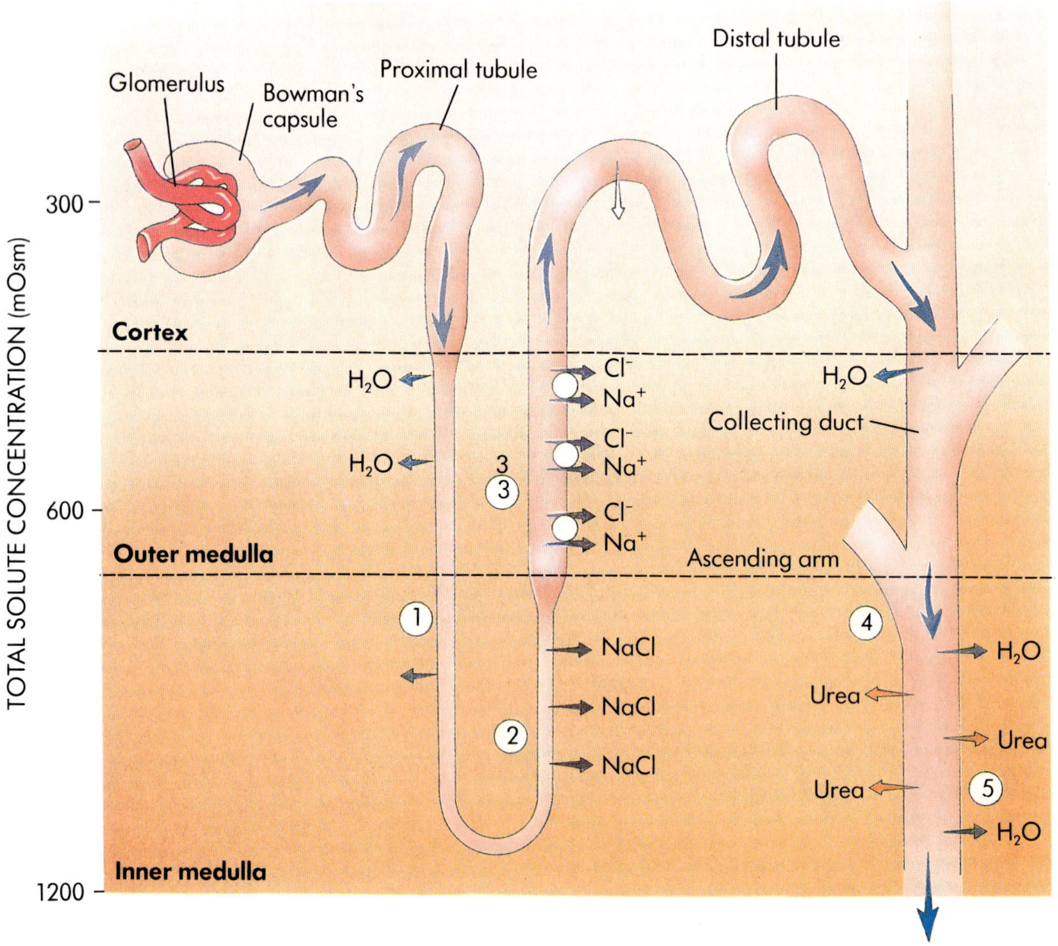

Flow of Materials in the Human Kidney
Figure 53–9

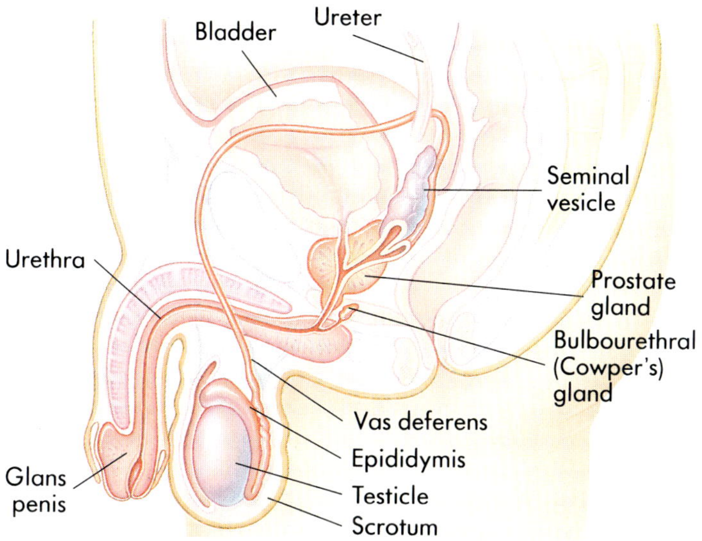

Male Reproductive Organs
Figure 54–6

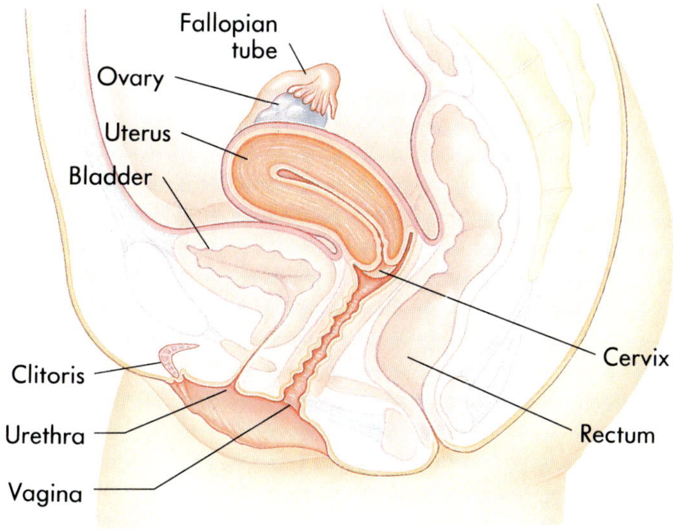

Female Reproductive Organs
Figure 54–11

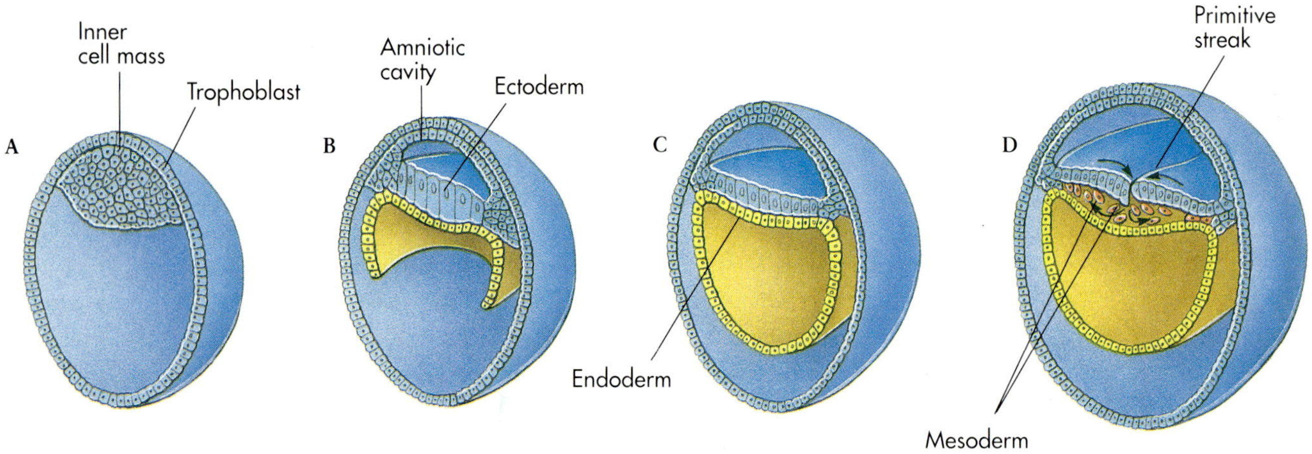

Mammalian Gastrulation
Figure 55–12